Improved Analytical Technologies for the Detection of Natural Toxins and Their Metabolites in Food

Improved Analytical Technologies for the Detection of Natural Toxins and Their Metabolites in Food

Editor

Veronica Maria Teresa Lattanzio

MDPI • Basel • Beijing • Wuhan • Barcelona • Belgrade • Manchester • Tokyo • Cluj • Tianjin

Editor
Veronica Maria Teresa Lattanzio
National Research Council of Italy
Italy

Editorial Office
MDPI
St. Alban-Anlage 66
4052 Basel, Switzerland

This is a reprint of articles from the Special Issue published online in the open access journal *Toxins* (ISSN 2072-6651) (available at: https://www.mdpi.com/journal/toxins/special_issues/technologies_detection_toxins).

For citation purposes, cite each article independently as indicated on the article page online and as indicated below:

LastName, A.A.; LastName, B.B.; LastName, C.C. Article Title. *Journal Name* **Year**, *Article Number*, Page Range.

ISBN 978-3-03936-920-1 (Hbk)
ISBN 978-3-03936-921-8 (PDF)

Contents

About the Editor

Veronica Maria Teresa Lattanzio Chemist. Senior Researcher at the Institute of Sciences of Food Production—National Research Council of Italy. Coordinator of the H2020 project "FOODSAFETY4EU: Multi-Stakeholder Platform for Food Safety in Europe". Her main research topic is the development and validation of analytical methods for mycotoxin and pesticides detection either based on mass spectrometry techniques or immunoassays, including organization of collaborative trials. She is a Member of the Working Group "Biotoxins" (CEN/TC 275 WG5 "Food Analysis–Horizontal Methods–Biotoxins), of the European Committee for Standardization. She collaborates with the International Atomic Energy Agency (IAEA) as an expert to provide individual and collective training related to testing chemical contamination for food safety. She also has specific expertise in identification, structural characterization, and determination of mycotoxin metabolites and modified mycotoxins; the study of the fate of mycotoxins during food processing; and the identification and characterization of biomarkers for the evaluation of animal and human exposure to mycotoxins.

toxins

MDPI

Editorial

Introduction to the Toxins Special Issue on Improved Analytical Technologies for the Detection of Natural Toxins and Their Metabolites in Food

Veronica Maria Teresa Lattanzio

Institute of Sciences of Food Production, National Research Council of Italy, Via Amendola, 122/O, 70126 Bari, Italy; veronica.lattanzio@ispa.cnr.it

Received: 15 July 2020; Accepted: 21 July 2020; Published: 22 July 2020

Food, by nature, is a biological substrate and is therefore capable of supporting the growth of microbials that are potential producers of toxic compounds. Natural toxins include mycotoxins, marine biotoxins, plant toxins, cyanogenic glycosides, and toxins occurring in poisonous mushrooms. Natural toxins pose not only a risk to both human and animal health, but also impact food security and nutrition by reducing people's access to healthy food. The tracking and detection of natural toxins in foods back to their source is a primary responsibility of food producers, distributors, handlers, and vendors. National authorities should conduct monitoring and ensure that levels of the most relevant natural toxins in food commodities comply with both national and international maximum levels or relevant toxicological thresholds.

This Special Issue of *Toxins* includes some recent advances in analytical methodologies for the detection of natural toxins in food commodities and biological fluids. The collected contributions are relevant to two main analytical approaches, namely advanced liquid chromatography–tandem mass spectrometry (LC-MS/MS) applications for the high sensitive and selective detection of emerging contaminants, and approaches for rapid and cost effective toxin detection based on immunoassays and cell-based assays.

The phytotherapeutic and nutritional use of plants and herbal-based products have increased in popularity over the last few years, however the improper usage of plants or unawareness of plant toxicity may lead to intoxication cases. Nowadays, highly sensitive and selective LC-MS/MS techniques represent a relatively easy to access and powerful tool to target plant toxins at low levels, and as a consequence, contaminants such as pyrrolizidine alkaloids and cardiac glycosides have surfaced as an issue of relevance in food safety.

A method for the determination of pyrrolizidine alkaloids in honey has been set up by Hungerford et al. [1]. The study includes the results of a large survey on honey samples from Queensland (Australia), providing a picture of pyrrolizidine alkaloids incidence and levels and a possible correlation with their plant origin. Malysheva et al. proposed a new LC-MS/MS method for the quantification of cardiac glycosides in edible herbs and spices and, complementary to this, in human urine [2]. The results of the validation indicated that the developed method is able to detect and quantify the target cardiac glycosides at low levels making it suitable for application in food safety controls and analysis of human biological fluids to unravel the level of exposure to plant toxins.

Among the readily available in vitro methods, the functional assay known as the cell-based assay, which uses a neuroblastoma (N2a) cell line, is widely applied for routine surveillance of marine toxins. In their study, Viallon et al. optimized several key parameters of the cell-based assay, improving assay implementation and reliable toxin detection, making the final protocol suitable for future standardization and interlaborarory comparisons [3].

The overall performances of immunoassays strongly depend on the quality of selected immunoreagents, first of all, of the antibody. For this reason, the development and testing of new antibodies continues to warrant considerable efforts in current research. Maragos et al. proposed a new competitive enzyme-linked immunosorbent assay (ELISA) to screen for citreoviridin and its geometric isomer, iso-citreoviridin, in white rice based upon the isolation of two novel monoclonal antibodies [4]. The developed assays were relatively tolerant to methanol and acetonitrile and provided adequate analytical performances such as sensitivity, accuracy and precision, allowing for the screening of citreoviridin and iso-citreoviridin at levels that are toxicologically relevant. Studies by Bever et al. deal with the generation of novel monoclonal antibodies and their application in a competitive ELISA and a lateral flow immunoassay (LFIA) for the detection of amatoxins, lethal toxins found in a variety of mushroom species. The developed assays are intended for the screening of wild mushrooms (ELISA) and urine samples (LFIA), thus providing complementary tools for evaluating amatoxins occurrence and geographic distribution as well as directly determining amatoxin exposure of humans and animals [5,6].

When applying rapid screening tests for compliance testing, their fitness for purpose needs to be demonstrated through a validation study. The manuscript by Pecorelli et al. provides insights about the process of evaluating and comparing the performance profile of rapid methods currently applied for AFM1 screening in milk [7]. Addressing EU official guidelines, the analytical performances of the strip test and ELISA-based methods were evaluated, including verification of method performances through long-term quality control measurements and comparison with the AOAC reference method.

Antibodies as well as alternative receptors, such as aptamers, can be used in several biosensing platforms. The review by Zhuheng Li et al. updates the construction strategies of electrochemical biosensors such as immunosensors and aptasensors for the cost-effective determination of microbial toxins including bacterial toxins, fungal toxins and algal toxins [8]. The paper summarizes the roles of 2D nanomaterials and their nanocomposites in the configuration of electrochemical biosensors, while also discussing advantages, major challenges and perspectives of these electrochemical biosensors for future commercialization.

Overall, the papers included in this Special Issue have contributed to advance the state-of-art of analytical methods for the detection of natural toxins. Furthermore, a part of the published studies focused on emerging or less investigated toxins, thus providing the scientific community with new tools and/or data supporting a better understanding of related food safety issues.

Funding: This research received no external funding.

Acknowledgments: The editor wishes to aknowledge all authors who contributed to this Special Issue and the expert peer reviewers performing careful and rigorous evaluations. The valuable contribution and editorial support of the MDPI management team and staff is also aknowledged.

Conflicts of Interest: The author declares no conflict of interest.

References

1. Hungerford, N.L.; Carter, S.J.; Anuj, S.R.; Tan, B.L.L.; Hnatko, D.; Martin, C.L.; Sharma, E.; Yin, M.; Nguyen, T.T.P.; Melksham, K.J.; et al. Analysis of Pyrrolizidine Alkaloids in Queensland Honey: Using Low Temperature Chromatography to Resolve Stereoisomers and Identify Botanical Sources by UHPLC-MS/MS. *Toxins* **2019**, *11*, 726. [CrossRef] [PubMed]
2. Malysheva, S.V.; Mulder, P.P.J.; Masquelier, J. Development and Validation of a UHPLC-ESI-MS/MS Method for Quantification of Oleandrin and Other Cardiac Glycosides and Evaluation of Their Levels in Herbs and Spices from the Belgian Market. *Toxins* **2020**, *12*, 243. [CrossRef] [PubMed]
3. Viallon, J.; Chinain, M.; Taiana Darius, H. Revisiting the Neuroblastoma Cell-Based Assay (CBA-N2a) for the Improved Detection of Marine Toxins Active on Voltage Gated Sodium Channels (VGSCs). *Toxins* **2020**, *12*, 281. [CrossRef] [PubMed]
4. Maragos, C.M.; Uchiyama, Y.; Kobayashi, N.; Kominato, F.; Sugita-Konishi, Y. Development and Characterization of Monoclonal Antibodies for the Mycotoxin Citreoviridin. *Toxins* **2019**, *11*, 630. [CrossRef] [PubMed]

5. Bever, C.S.; Hnasko, R.M.; Cheng, L.W.; Stanker, L.H. A Rapid Extraction Method Combined with a Monoclonal Antibody-Based Immunoassay for the Detection of Amatoxins. *Toxins* **2019**, *11*, 724. [CrossRef] [PubMed]
6. Bever, C.S.; Swanson, K.D.; Hamelin, E.I.; Filigenzi, M.; Poppenga, R.H.; Kaae, J.; Cheng, L.V.; Stanker, L.H. Rapid, Sensitive, and Accurate Point-of-Care Detection of Lethal Amatoxins in Urine. *Toxins* **2020**, *12*, 123. [CrossRef] [PubMed]
7. Pecorelli, I.; Guarducci, N.; von Holst, C.; Bibi, R.; Pascale, M.; Ciasca, B.; Logrieco, A.F.; Lattanzio, V.M.T. Critical Comparison of Analytical Performances of Two Immunoassay Methods for Rapid Detection of Aflatoxin M1 in Milk. *Toxins* **2020**, *12*, 270. [CrossRef] [PubMed]
8. Li, Z.; Li, X.; Jian, M.; Selale Geleta, G.; Wang, Z. Two-Dimensional Layered Nanomaterial-Based Electrochemical Biosensors for Detecting Microbial Toxins. *Toxins* **2020**, *12*, 20. [CrossRef] [PubMed]

toxins

MDPI

Article

Analysis of Pyrrolizidine Alkaloids in Queensland Honey: Using Low Temperature Chromatography to Resolve Stereoisomers and Identify Botanical Sources by UHPLC-MS/MS

Natasha L. Hungerford [1,*], Steve J. Carter [2], Shalona R. Anuj [2], Benjamin L. L. Tan [2], Darina Hnatko [2], Christopher L. Martin [1], Elipsha Sharma [1], Mukan Yin [1], Thao T. P. Nguyen [1], Kevin J. Melksham [2] and Mary T. Fletcher [1,*]

1 Queensland Alliance for Agriculture and Food Innovation (QAAFI), The University of Queensland, Brisbane, QLD 4072, Australia; christopherlancemartin@gmail.com (C.L.M.); e.sharma@uq.net.au (E.S.); mukan.yin@uq.net.au (M.Y.); tran.nguyen7@uq.net.au (T.T.P.N.)

2 Forensic and Scientific Services, Queensland Health, Brisbane, QLD 4108, Australia; steve.carter@health.qld.gov.au (S.J.C.); shalona.anuj@health.qld.gov.au (S.R.A.); benjamin.tan@health.qld.gov.au (B.L.L.T.); Darina.Hnatko@health.qld.gov.au (D.H.); Kevin.Melksham@health.qld.gov.au (K.J.M.)

* Correspondence: n.hungerford@uq.edu.au (N.L.H.); mary.fletcher@uq.edu.au (M.T.F.)

Received: 25 October 2019; Accepted: 26 November 2019; Published: 11 December 2019

Abstract: Pyrrolizidine alkaloids (PAs) are a diverse group of plant secondary metabolites with known varied toxicity. Consumption of 1,2-unsaturated PAs has been linked to acute and chronic liver damage, carcinogenicity and death, in livestock and humans, making their presence in food of concern to food regulators in Australia and internationally. In this survey, honey samples sourced from markets and shops in Queensland (Australia), were analysed by high-resolution Orbitrap UHPLC-MS/MS for 30 common PAs. Relationships between the occurrence of pyrrolizidine alkaloids and the botanical origin of the honey are essential as pyrrolizidine alkaloid contamination at up to 3300 ng/g were detected. In this study, the predominant alkaloids detected were isomeric PAs, lycopsamine, indicine and intermedine, exhibiting identical MS/MS spectra, along with lesser amounts of each of their *N*-oxides. Crucially, chromatographic UHPLC conditions were optimised by operation at low temperature (5 °C) to resolve these key isomeric PAs. Such separation of these isomers by UHPLC, enabled the relative proportions of these PAs present in honey to be compared to alkaloid levels in suspect source plants. Overall plant pyrrolizidine alkaloid profiles were compared to those found in honey samples to help identify the most important plants responsible for honey contamination. The native Australian vines of *Parsonsia* spp. are proposed as a likely contributor to high levels of lycopsamine in many of the honeys surveyed. Botanical origin information such as this, gained via low temperature chromatographic resolution of isomeric PAs, will be very valuable in identifying region of origin for honey samples.

Keywords: LC-MS; pyrrolizidine alkaloid; honey; *Parsonsia straminea*; lycopsamine; indicine; *Heliotropium amplexicaule*

Key Contribution: This study demonstrated the occurrence of indicine and the isomeric lycopsamine as predominant pyrrolizidine alkaloids in Australian honey; and established a HRAM LC-MS/MS method that chromatographically separated these stereoisomers; enabling the investigation of botanical origin of honey contamination by the respective alkaloids.

1. Introduction

Pyrrolizidine alkaloids (PAs) are secondary metabolites that comprise more than 600 compounds, typified by a pyrrolizidine base with one or more ester linkages. The 1,2-unsaturated PAs are toxic to animals and humans, causing acute and chronic liver and lung damage or cancer and are mainly produced by flowering plant species belonging to the families Asteraceae (Compositae, tribes Senecioneae and Eupatorieae), Fabaceae (Crotolaria, Chromolaena, Lotonis), Apocynaceae (Echiteae) and Boraginaceae [1], estimated to represent 3% of the Earth's flowering plants [2]. Plants containing pyrrolizidine alkaloids are globally distributed and PAs provide a chemical defence for plants against herbivores. PA biosynthesis has been found to be dependent on many factors, with implications for plant/animal/insect interactions [3]. Various food products can contain toxic PAs either directly from plant origin (certain herbs, herbal medicines) or indirectly through natural transfer from floral nectar and pollen (e.g., some honey, pollen dietary supplements) or inadvertent cross-contamination (e.g., grains, herbs, teas) [4]. The Australian provisional tolerable daily intake of pyrrolizidine alkaloids is 1 μg/kg Bw/day [5], whilst the recommended European accepted intake is 150 times lower at 0.007 μg/kg Bw/day [6–8], although this has been recently revised [9,10].

Several studies have described the presence of toxic pyrrolizidine alkaloids (PAs) in honey produced by *Apis mellifera*. Internationally it has been reported that such toxins can be found in honey due to transfer by bees of pollen/nectar from certain flowers, including *Heliotropium*, *Crotolaria*, *Echium* and *Senecio* species. Concern has been raised as to the extent of contamination in Australian honey [11,12] but their presence in Queensland honey has not previously been examined. Previously, investigations into the presence of PAs in Australian honey have concentrated on the introduced pest plant *Echium plantagineum* L. (Paterson's curse) as the major likely PA source [11–13]. However, rigorous eradication and biological control programs in recent decades have decreased the prevalence of this weed in Australia [14]. Diverse PA containing plant species occur in Australian pastures and have intoxicated grazing horses, cattle, sheep or pigs and poultry fed contaminated grains [15,16] and include both native and introduced *Crotalaria, Heliotopium* and *Senecio* species [17]. Additionally, native Australian *Parsonsia* species grow in rainforests and on the margins of rainforest/eucalypt forest and contain PAs known to be sequestered by butterflies [18–20]. *Parsonsia* species spread from tropical and subtropical Asia to Australia and the south-west Pacific. *Parsonsia straminea* is native to Queensland and New South Wales [21] but there have been no field reports of livestock poisonings [17]. The distribution of these PA containing plants varies throughout the country and different alkaloids would be expected in honey from tropical/sub-tropical Queensland regions as compared to honey originating from southern temperate states. Given that honey represents a significant food source of human exposure to PAs [10], identification of plant PA sources to reduce this exposure is crucial.

In this study of market honey, samples have been identified with alkaloid profiles that appeared to be consistent with a number of PA containing plant species present within the Australian environment This study examines the presence of PAs in a market survey of honey purchased in Queensland with the aim to assess any food safety concern for the consumer of honey and to correlate PAs identified with previously unsuspected plant sources of these alkaloids.

2. Results and Discussion

2.1. Pyrrolizidine Analysis Method Validation

Pyrrolizidine alkaloid levels in honey/plant material were quantitated by HRAM UHPLC-MS/MS analysis against 30 certified PA standards, through comparison of the precursor parent ion intensity (Table 1) to the standard curves, with squared correlation coefficients (R^2) typically in the range of 0.9932–0.9997. The honey analysis method was validated according to the National Association of Testing Authorities (NATA) guidance document [22]. The method was validated in blank honey, based on results for 10 spiked samples, giving Limits of Reporting (LORs) of 5 ng/g for individual PAs (Table 2). The uncertainties given are at the 95% confidence level as required by NATA [22,23].

Note that for multiresidue analyses at these levels, a default standard uncertainty of ±25% RSD at the 95% confidence level is routinely applied by the authors and is used unless there is evidence that the actual uncertainty is greater than this value. Erucifoline, erucifoline *N*-oxide, jacobine *N*-oxide and seneciphylline *N*-oxide consistently gave low recoveries, resulting in high calculated standard uncertainty (% RSD) for these PAs, but which are not unusual for analyses at these levels. The EU recommends ±50% RSD unless it is demonstrated to be a bigger value [24]. The uncertainty values are calculated at the LOR because it is expected that this level will be the worst case scenario. A small peak was present in the blank for trichodesmine, explaining the higher LOD/LOR and increased uncertainty for this compound.

Table 1. Details of pyrrolizidine alkaloids used in the Orbitrap analysis of PA containing plants and honey, including formulae, retention times, precursor ions used for quantitation and confirmatory product ions.

Compound	Formula	Average R_T (min)	Precursor Ion (MH^+)	Product Ions					
Echimidine (**7**)	$C_{20}H_{31}NO_7$	12.86	398.2173	120.0809	83.0497	55.0550			
Echimidine *N*-oxide	$C_{20}H_{31}NO_8$	12.80	414.2122	396.2004	352.1745	254.1379	220.1326	137.0833	
Erucifoline	$C_{18}H_{23}NO_6$	6.23	350.1598	322.1642	220.1329	164.1066	138.0911	120.0807	
Erucifoline *N*-oxide	$C_{18}H_{23}NO_7$	8.03	366.1547	278.1386	218.1172	164.1067	136.0756	119.0729	
Europine	$C_{16}H_{27}NO_6$	6.97	330.1911	254.1385	156.1019	138.0914	120.0810	96.0812	
Europine *N*-oxide	$C_{16}H_{27}NO_7$	7.86	346.1860	328.1743	270.1328	256.1172	172.0964	155.0937	
Helioamplexine	$C_{16}H_{27}NO_5$	9.21	314.1962	156.1017	138.0913	120.0808	94.0655		
Helioamplexine *N*-oxide	$C_{16}H_{27}NO_6$	10.42	330.1911	172.0966	155.0938	138.0913	111.0913	94.0653	
Heliotrine	$C_{16}H_{27}NO_5$	9.56	314.1962	156.1017	138.0913	120.0808	94.0655		
Heliotrine *N*-oxide	$C_{16}H_{27}NO_6$	10.46	330.1911	172.0966	155.0938	138.0913	111.0913	94.0653	
Indicine (**2**)	$C_{15}H_{25}NO_5$	6.67	300.1806	156.1019	138.0914	120.0810	94.0656	82.0657	
Indicine *N*-oxide (**5**) and intermedine *N*-oxide (**4**) (n.r.) [a]	$C_{15}H_{25}NO_6$	8.20	316.1755	226.1437	172.0968	155.0941	138.0914	111.0682	94.0656
Intermedine (**1**)	$C_{15}H_{25}NO_5$	6.26	300.1806	210.1488	156.1019	138.0914	120.0810	94.0656	
Jacobine	$C_{18}H_{25}NO_6$	6.56	352.1755	308.1485	280.1539	262.1432	234.1483	155.1063	
Jacobine *N*-oxide	$C_{18}H_{25}NO_7$	7.91	368.1704	296.1485	190.1222	139.0989	121.0885	120.0807	
Lasiocarpine	$C_{21}H_{33}NO_7$	14.92	412.2330	238.1435	156.1020	138.0914	120.0810	94.0656	
Lasiocarpine *N*-oxide	$C_{21}H_{33}NO_8$	16.14	428.2279	410.2168	352.1746	328.1753	254.1384	220.1333	137.0835
Lycopsamine (**3**)	$C_{15}H_{25}NO_5$	6.80	300.1806	156.1017	138.0914	120.0808	94.0655		
Lycopsamine *N*-oxide (**6**)	$C_{15}H_{25}NO_6$	8.65	316.1755	172.0964	155.0937	138.0911	136.0755	94.0654	
Monocrotaline	$C_{16}H_{23}NO_6$	2.88	326.1598	280.1548	237.1354				
Monocrotaline *N*-oxide	$C_{16}H_{23}NO_7$	7.19	342.1547	314.1590	296.1487	236.1274	137.0833	119.0729	
Retrorsine	$C_{18}H_{25}NO_6$	9.13	352.1755	324.1802	138.0913	120.0808	94.0655		
Retrorsine *N*-oxide	$C_{18}H_{25}NO_7$	9.64	368.1704	220.1340	154.0862				
Senecionine	$C_{18}H_{25}NO_5$	11.13	336.1806	308.1864	120.0809				
Senecionine *N*-oxide	$C_{18}H_{25}NO_6$	11.83	352.1755	324.1825	220.1332				
Seneciphylline	$C_{18}H_{23}NO_5$	9.52	334.1649	306.1706	120.0811				
Seneciphylline *N*-oxide	$C_{18}H_{23}NO_6$	10.46	350.1598	322.1656	246.1495				
Senecivernine	$C_{18}H_{25}NO_5$	10.84	336.1806	308.1848	153.0907	138.0911	120.0807	94.0654	
Senecivernine *N*-oxide	$C_{18}H_{25}NO_6$	11.45	352.1755	324.1795	220.1327	154.0859	136.0755	120.0807	
Senkirkine	$C_{19}H_{27}NO_6$	13.62	366.1911	168.1020	150.0915				
Trichodesmine	$C_{18}H_{27}NO_6$	8.79	354.1911	308.1857	223.1203	222.1489	164.1071	121.0889	

[a] n.r. = not resolved.

Table 2. Method validation results for pyrrolizidine alkaloids in honey.

Display Name	Spiking Level (ng/g)	LOD (3s)	LOQ (9s)	LOR	Units	Recovery (%)	Calculated Standard Uncertainty (% RSD)
Echimidine (**7**)	5	0.7	2	5	ng/g	88	11
Echimidine *N*-oxide	5	0.5	2	5	ng/g	60	15
Erucifoline	5	0.7	2	5	ng/g	34	26
Erucifoline *N*-oxide	5	0.5	2	5	ng/g	26	31
Europine	5	1	2	5	ng/g	98	16
Europine *N*-oxide	5	0.5	1	5	ng/g	94	6
Heliotrine	5	0.5	0.7	5	ng/g	108	4
Heliotrine *N*-oxide	5	0.2	0.5	5	ng/g	83	2
Indicine	5	0.5	2	5	ng/g	98	8
Indicine *N*-oxide (**5**) and intermedine *N*-oxide (**4**) as indicine *N*-oxide (**5**)	5	0.5	1	5	ng/g	78	7
Intermedine (**1**)	5	0.5	2	5	ng/g	94	8
Jacobine	5	2	5	5	ng/g	61	30
Jacobine *N*-oxide	5	1	5	5	ng/g	36	51
Lasiocarpine	5	0.5	1	5	ng/g	82	6
Lasiocarpine *N*-oxide	5	0.5	0.7	5	ng/g	79	6
Lycopsamine (**3**)	5	0.2	0.7	5	ng/g	95	4
Lycopsamine *N*-oxide (**6**)	5	0.7	2	5	ng/g	81	14
Monocrotaline	5	2	5	5	ng/g	89	24
Monocrotaline *N*-oxide	5	0.7	2	5	ng/g	89	13
Retrorsine	5	1	2	5	ng/g	96	13
Retrorsine *N*-oxide	5	0.7	2	5	ng/g	58	22
Senecionine	5	0.7	2	5	ng/g	90	11
Senecionine *N*-oxide	5	0.5	2	5	ng/g	58	16
Seneciphylline	5	0.7	2	5	ng/g	70	15
Seneciphylline *N*-oxide	5	0.5	1	5	ng/g	16	36
Senecivernine	5	0.5	2	5	ng/g	86	9
Senecivernine *N*-oxide	5	0.7	2	5	ng/g	53	20
Senkirkine	5	0.5	0.7	5	ng/g	91	5
Trichodesmine	5	3	7	7	ng/g	78	69

2.2. *Alkaloid Levels Measured in Honey*

Honeys purchased in supermarkets, health food shops, and from individual commercial/small-scale producers were analysed and calculated to contain pyrrolizidine alkaloids levels between <LOR (i.e., below limit of reporting) to ≈3300 ng/g of honey.

Figure 1 summarises the results in a histogram, with single PA test results below the limit of reporting (5 ng/g) set equal to zero. PAs were detected in 84% of the honey samples examined (n = 465). Notably the mean total PA level of PA-positive samples (280 ng/g) was greater than the median (97 ng/g), indicating that the distribution was skewed, with a prevalence of low values (Figure 1, histogram). Whilst the prevalence of low values is reassuring, the overall distribution of total PA concentration is wide, ranging from <LOR to ≈3300 ng/g.

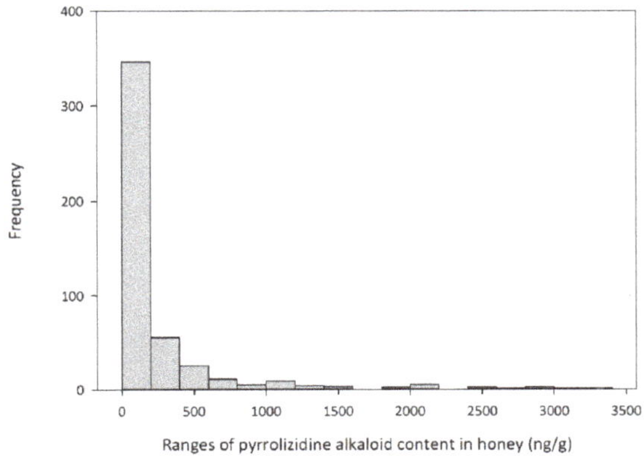

Figure 1. Histogram showing the frequency of total pyrrolizidine alkaloid concentrations in honey samples (n = 465) analysed against all 30 pyrrolizidine alkaloid standards (and isolated helioamplexine [25]).

2.3. LC-MS/MS Separation of Alkaloids

Analysis revealed that the individual PA pattern detected by the LC-MS/MS analysis of honeys was characterised almost exclusively by lycopsamine-type PAs. In this study the lycopsamine-type PAs were represented by standards intermedine (**1**), indicine (**2**) and lycopsamine (**3**) (Figure 2). These diastereomeric PAs cannot be distinguished based on their MS/MS spectra [26], and Figure 3 shows the identical mass spectra obtained for standards intermedine (**1**), indicine (**2**) and lycopsamine (**3**) by our described HRAMS method. Given the diastereomeric nature of these alkaloids all parent MH^+ ions and fragment ions are identical, even with HRAMS. Separation based on retention time (RT) was therefore necessary in order to ascertain the botanical origin of PA contamination in these honeys. In most previous studies of PAs in honey, lycopsamine-type PAs were reported as the sum of unresolved stereoisomers, (including indicine (**2**), intermedine (**1**) and/or lycopsamine (**3**), and even the less common rinderine and echinatine) [27–30] or partially resolved stereoisomers [8,31–33]. Under our initial UHPLC conditions, with a column oven temperature of 40 °C, intermedine (**1**) eluted separately first, but indicine (**2**) and lycopsamine (**3**) co-eluted from the Kinetex XB-C18 UHPLC column. Notably the combined indicine/lycopsamine (**2/3**) peak represented 75% of the alkaloids present in Queensland honey. As these two alkaloids originate from distinctly different PA plant sources, our aim was to be able to separately quantify the levels of each of these PAs in honey to enable the major plant source of PA contamination to be identified.

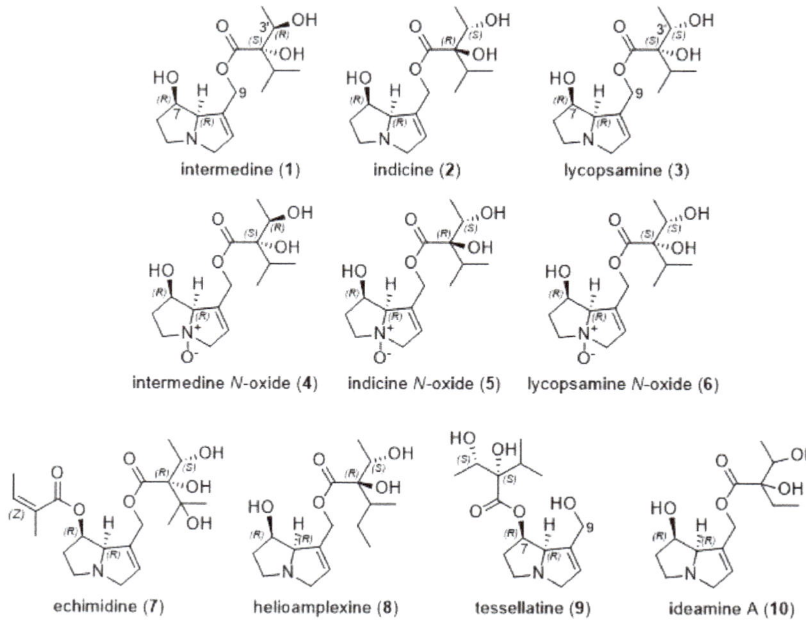

Figure 2. Structures of predominant PAs (**1–8**) observed in honey (n = 465), together with structures of minor components tentatively observed in *Parsonsia straminea* (**9–10**).

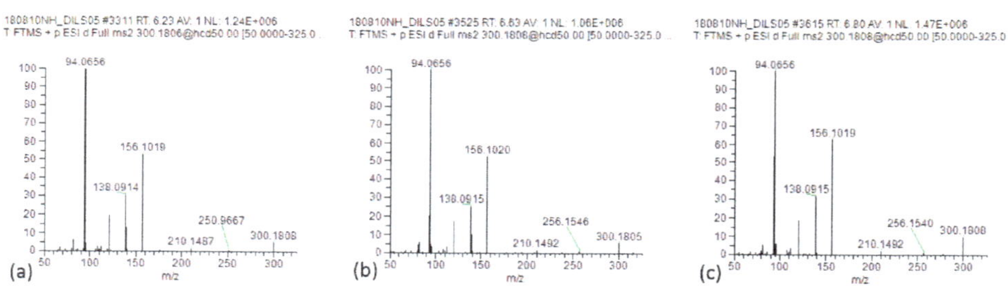

Figure 3. Stereoisomeric pyrrolizidine alkaloids (**a**) intermedine (**1**), (**b**) indicine (**2**), and (**c**) lycopsamine (**3**), with identical high resolution accurate mass spectra.

Ultimately, separation of indicine/lycopsamine (**2/3**) was achieved by simply adjusting the column temperature to 5 °C. A more complicated 'multiple heart-cutting two dimensional chromatography' method has previously been reported for the resolution of multiple PA isomer pairs [34], but in our hands the simple gradient elution at 5 °C was sufficient to achieve our desired resolution of indicine/lycopsamine (**2/3**). Under these conditions, of the 30 PAs and PA-NOs all were resolved based on retention time or mass fragmentation of the MS/MS except for intermedine *N*-oxide (**4**) and indicine *N*-oxide (**5**) which displayed identical RT and MS/MS (Figure 4). In plants where these *N*-oxides (**4**) and (**5**) are prevalent, the *N*-oxides could be distinguished by reduction to the corresponding parent alkaloid (**2/3**) which were resolved by RT under the described conditions.

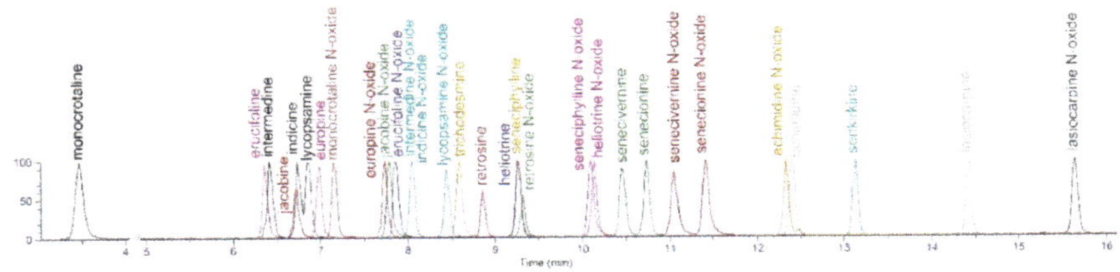

Figure 4. Extracted ion chromatogram of the 30 pyrrolizidine alkaloid calibration standards, illustrating the separation obtained under the UHPLC method, with column temperature of 5 °C.

2.4. Predominant Alkaloids Present in Queensland Honeys

Analysis of all 465 honeys under our optimised LC-MS/MS conditions revealed that the predominant pyrrolizidine alkaloid present in our Queensland honey samples was lycopsamine (**3**), which represented approximately 51% of the measured alkaloid content, followed by indicine (**2**) at 24%, lycopsamine *N*-oxide (**6**) at 9%, intermedine (**1**) at 6% and echimidine (**7**) at 3% (Figure 5). Even though we did not resolve intermedine *N*-oxide (**4**) and indicine *N*-oxide (**5**), the identity of the minor *N*-oxide in individual honey samples was inferred by the presence of the co-occurring parent alkaloid (either intermedine (**1**) or indicine (**2**)).

In individual honeys, lycopsamine (**3**) was detected at up to ≈3100 ng/g, indicine (**2**) at up to 1700 ng/g, with the highest total PA content in any individual honey of ≈3300 ng/g which contained mainly a mixture of lycopsamine (**3**) and lycopsamine *N*-oxide (**6**).

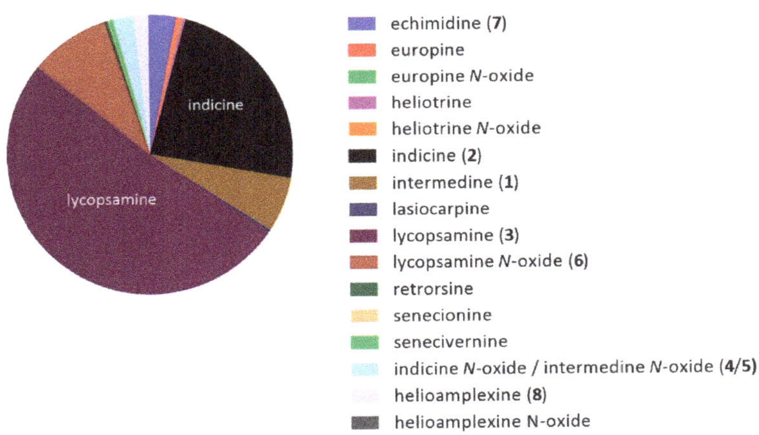

Figure 5. Total amount of each pyrrolizidine alkaloid detected against the 30 PA standards.

Figure 6 shows a Tukey box and whisker plot of the pyrrolizidine alkaloids detected in honeys (n = 465), showing the distribution of each PA concentration, for positive samples only. The largest variation was observed for lycopsamine (**3**), indicine (**2**) and lycopsamine *N*-oxide (**6**). In honeys where lycopsamine (**3**) and its *N*-oxide (**6**) were abundant these were generally the dominant PAs (>90% of PAs detected).

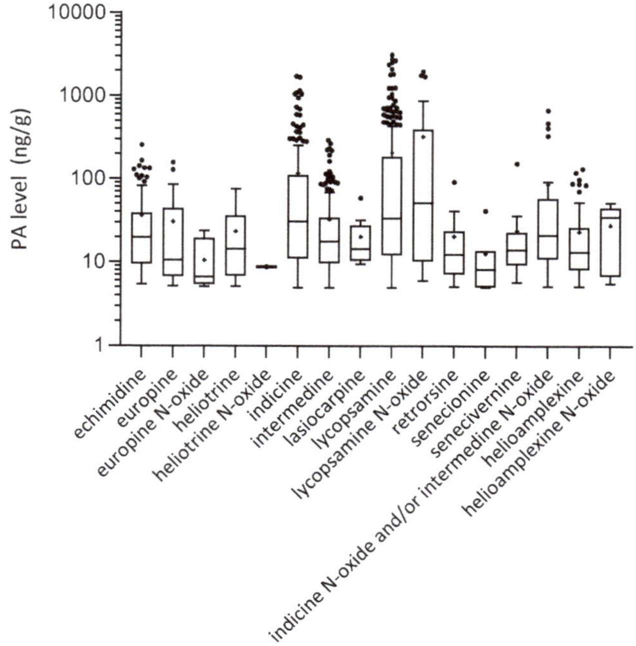

Figure 6. Tukey box and whisker plot of distribution of each pyrrolizidine alkaloid detected in honey (n = 465) (includes results >5 ng/g only).

Similarly, in honeys where indicine (**2**) and its *N*-oxide (**5**) were abundant these were generally the dominant PAs (>68% of PAs detected). In order to explain the relative predominance of these diastereomeric PAs in different honeys, it was clear that we had to identify two main and distinctly different PA plant sources.

2.5. *Plant Sources of Indicine* (**2**) *in Honey*

An examination of the locally abundant weed *Heliotropium amplexicaule* (Blue heliotrope) by our LC-MS/MS method revealed that indicine (**2**) and indicine *N*-oxide (**5**) were the predominant pyrrolizidine alkaloids in this plant, and examination of more minor components including the newly identified helioamplexine (**8**) provided a unique fingerprint in the HRAM LC-MS/MS profile [25]. Interrogation of the pyrrolizidine alkaloid profile from market honey samples with high amounts of indicine (**2**), demonstrated that there was strong correlation between the honey PA profile and the *H. amplexicaule* plant alkaloid profile. The presence of both major and minor *H. amplexicaule* alkaloids in this honey provided strong evidence that this plant represented the floral source for this alkaloid contamination [25].

2.6. *Plant Sources of Lycopsamine* (**3**) *in Honey*

We similarly sought to understand the source of lycopsamine (**3**) (and its *N*-oxide (**6**)), the major PA observed in Queensland honey. Examination of the PAs co-occurring with lycospamine (**3**) and lycopsamine *N*-oxide (**6**), in the source plant would enable us to establish a unique floral PA fingerprint that could be correlated with PAs observed in honey. In past studies, *Echium plantagineum* L. (Paterson's curse) has been named as the source of lycopsamine (**3**) in Australian honey [35], despite the fact lycopsamine (**3**) is usually only a minor alkaloid in *Echium* spp. [11,27,36,37]. In fact, a previous European study noted the presence of high amounts of lycopsamine (**3**) (607 ng/g) compared to low amounts of echimidine (7) (15 ng/g) in imported Australian honeys, and postulated an unknown plant source as a possible interpretation [27]. Indeed our analysis of *E. plantagineum* revealed that after Zn reduction echimidine (7) and echiumine were the dominant PAs, with both lycopsamine (3) and intermedine (1) present in much lower quantities. Clearly *E. plantagineum* is not the major source of lycopsamine (3) seen in our Queensland honeys, which is also consistent with the more temperate distribution of this species within Australia [38]. Other species/genera known internationally to contain lycopsamine (3) (and intermedine (1)) include *Anchusa off., Borago off., Lithospermum* spp., and *Symphytum* spp., and *Eupatorium* spp. [39], and are generally not geographically distributed within Australia [40]. They can logically be excluded as potential lycopsamine (3) floral sources.

When considering PA species which are known to be prevalent in Queensland, both *Ageratum* and *Aminscka* spp. have been reported to contain lycopsamine (3). *Ageratum conyzoides* for example has been reported to contain lycopsamine (3) and echinatine [41,42] or lycopsamine (3) and 3′-*O*-acetyllycopsamine [43]. A targeted screen by Avula reported lycopsamine (3) and its *N*-oxide (6) as the two major PAs, together with minor amounts of dihydrolycopsamine, dihydrolycopsamine *N*-oxide and echinatine [1,44]. The closely related *Ageratum houstonianum* is locally abundant in Queensland, and our analysis of Zn reduced plant extract revealed the predominance of retrohoustine, heliohoustine and tentatively echinatine (ratio 2.7:1.7:1 respectively), with much lower amounts of lycopsamine (3) and intermedine (1) (data not shown). This result is consistent with analysis of this same species from Mexico that showed that lycopsamine (3) was not the predominant pyrrolizidine alkaloid present with three other pyrrolizidine alkaloids (retrohoustine, heliohoustine and isoretrohoustine) isolated in greater amounts than lycopsamine (3) [45]. Lycopsamine (3) and intermedine (1) have also been identified in *Amscinckia* spp. [46], with NMR analysis revealing the relative proportion of intermedine (1) to lycopsamine (3) varied from roughly 2:1 to 1:2 in *A. intermedia, A. hispida,* and *A. lycopsoides. Amsinckia* spp. are however regionally controlled as noxious weeds in Australia, and not likely to be a widely abundant PA sources in Queensland. The invasive aquatic weed *Gymnocoronis spilanthoides* has been recently been shown [47] to contain predominantly lycopsamine (3) followed by intermedine (1), however, this species is also controlled by government eradication programs. None of these plant species matched either the predominant lycopsamine (3) profile observed in our Queensland sourced honey or the regional abundance of plant species.

Historically lycopsamine (3) was identified in the hair pencil of Australian danaid butterflies in Queensland in a region where *Amsinckia* plants are rare [48]. An examination of the native vines *Parsonsia straminea* (family Apocynaceae) and *Parsonsia eucalyptophylla*, by these authors revealed the presence of lycopsamine (3) and intermedine/indicine (1 or 2), and acetyl derivatives. As native *Parsonsia* species occur widely in Queensland this species was deduced as the source of lycopsamine (3) in danaid butterflies [49]. Lycopsamine-type PAs have been identified in a number of species in Apocynaceae [50].

Interestingly, in a study of butterfly food plants, a comparison of *Parsonsia straminea* flowers revealed the ratio of lycopsamine *N*-oxide (6) to intermedine *N*-oxide (4) to other alkaloids of 98:1:1. By contrast, *Ageratum sp.* gave a predominance of two M+ 269 isomers compared to lycopsamine (3) (45:48:1) [51]. Evidently, lycopsamine (3) and intermedine (1) and their *N*-oxides are present in a wide variety of plant species, but we sought to identify an origin for the almost exclusive predominance of lycopsamine (3) (and its *N*-oxide (6)) and these literature reports of *Parsonsia* provided the best clue.

2.7. Pyrrolizidine Alkaloids Determined in Parsonsia Vines

Local *Parsonsia straminea* (Qld Herbarium ID AQ522465) was collected and re-examined for PA content using our described HRAM LC-MS/MS method. The plant pyrrolizidine alkaloids were present primarily as the *N*-oxides (96% in the leaves and stems, 99% in the pods, 93% in the nectar and 80% in the pollen). The plant pyrrolizidine alkaloids were analysed with and without reduction by Zn to enable comparison with the honey alkaloids (primarily free alkaloids) as previously observed [25,52]. The SCX SPE methodology was previously demonstrated to be suitable for plant extracts [36]. The investigations aimed to determine for the first time whether and to what extent PAs found in honey are sourced from *Parsonsia straminea* (or closely-related *Parsonsia* species, a number of which are widespread in coastal regions of eastern Australia [53]). High resolution accurate mass (HRAM) data, combined with RT comparison with pyrrolizidine alkaloids standards enabled identification of the major pyrrolizidine alkaloids in *P. straminea* (Table 3).

Table 3. High resolution accurate mass (HRAM) data for pyrrolizidine alkaloids in *P. straminea* identified by comparison with PA standards.

Alkaloid	Typical R_T (min)	Molecular Ion Formula	Calculated $[M+H]^+$	Observed $[M+H]^+$	Mass Spectral Data (Rel. Int. %)
Intermedine (**1**)	6.27	$[C_{15}H_{25}NO_5 + H]^+$	300.1805	300.1804	300.1804 (6) 256.1539 (1) 156.1019 (53) 139.0992 (11) 138.0914 (31) 120.0809 (18) 112.0759 (2) 108.0811 (2) 96.0812 (4) 95.0733 (4) 94.0656 (100) 82.0657 (4)
Lycopsamine (**3**)	6.80	$[C_{15}H_{25}NO_5 + H]^+$	300.1805	300.1805	300.1805 (9) 256.1539 (0.3) 156.1020 (59) 139.0993 (13) 138.0915 (31) 120.0811 (19) 112.0760 (3) 108.0813 (1) 96.0813 (5) 95.0734 (5) 94.0656 (100) 82.0657 (6)
Intermedine *N*-oxide (**4**)	8.19	$[C_{15}H_{25}NO_6 + H]^+$	316.1755	316.1752	316.1752 (41) 298.1631 (1) 272.1491 (2) 172.0967 (100) 155.0940 (18) 154.0862 (8) 138.0914 (55) 137.0837 (6) 136.0757 (15) 112.0759 (5) 111.0682 (12) 94.0655 (19) 93.0577 (13) 82.0419 (3)
Lycopsamine *N*-oxide (**6**)	8.64	$[C_{15}H_{25}NO_6 + H]^+$	316.1755	316.1752	316.1752 (46) 298.1653 (1) 272.1491 (2) 172.0967 (100) 155.0940 (17) 154.0862 (10) 138.0913 (63) 137.0835 (6) 136.0758 (17) 112.0759 (5) 111.0681 (13) 94.0655 (22) 93.0577 (16) 82.0419 (4)

In the *P. straminea* nectar, the ratio of lycopsamine (**3**) and its *N*-oxide (**6**) to intermedine (**1**) and its *N*-oxide (**4**) was >45–50:1, in the flowers it was 78:1, in anthers/pollen >50:1, in the pods it was >50:1, whilst in the leaves, ~3:1.

Minor peaks after reduction were tentatively identified by analysis of the HRAM data (Table 4) and corresponded to tessellatine (**9**) or isomer (a C7 isomer, found 300.1801, calculated for $C_{15}H_{25}NO_5$+H^+: 300.1805), a further C9 lycopsamine isomer (found 300.1803, calculated for $C_{15}H_{25}NO_5$+H^+: 300.1805), 3′-*O*-acetyllycopsamine (found 342.1905, calculated for $C_{17}H_{28}NO_6$+H^+ 342.1917), 3′-*O*-acetylintermedine (found 342.1924, calculated for $C_{17}H_{28}NO_6$+H^+: 342.1917) and two helioamplexine isomers (found 314.1958 and 314.1958, calculated for $C_{16}H_{27}NO_5$+H^+: 314.1962). The corresponding *N*-oxides were found in the non-reduced plant extract. Tessellatine (**9**) has the same necic acid as lycopsamine (**3**) but is esterified at the C7 necine position rather than C9 as seen in lycopsamine (**3**). The C7 esterification is evidenced in the predominant (base peak) fragment ion *m/z* 156.1019 (calculated for $C_8H_{14}NO_2^+$ 156.1019) characteristic of C7 monoesters [54,55], which display much smaller peaks at *m/z* 138.0913, 120.0809 and 94.0656 than C9 monoesters lycopsamine (**3**)/indicine (**2**)/intermedine (**1**). The diastereomeric 3′-*O*-acetyllycopsamine and 3′-*O*-acetylintermedine exhibited a similar MS breakdown to that seen in 3′-*O*-angelylindicine [25], with a base peak of *m/z* 94.0655 and other typical peaks of C9 monoesters of retronecine, 156.1019, 138.0913 and 120.0809. In these acetyl compounds, the lack of a peak at *m/z* 198.1125 and the lack of a base peak at *m/z* 214.1074 in the corresponding *N*-oxides, excluded the 7-*O*-acetyl substitution pattern [1,56]. Similarly the two helioamplexine isomers had identical MS to that seen in helioamplexine (**8**) (the C-6′ homoanalogue of indicine) [25], and these components present in *P. straminea* which did not co-elute with helioamplexine were deduced to be the corresponding C-6′ homoanalogues of lycopsamine and intermedine.

Table 4. High resolution accurate mass (HRAM) data for pyrrolizidine alkaloids in *P. straminea* identified by analysis of fragmentation patterns, together with those identified in honey samples high in lycopsamine (identification in honey indicated by *).

Alkaloid	Typical R_T (min)	Molecular Ion Formula	Calculated $[M+H]^+$	Observed $[M+H]^+$	Mass Spectral Data (Rel. Int. %)
* Tessellatine (**9**) or isomer	6.34	$[C_{15}H_{25}NO_5 + H]^+$	300.1805	300.1801	300.1801 (5) 156.1019 (100) 139.0990 (1) 138.0913 (6) 120.0809 (2) 112.0759 (4) 108.0810 (6) 96.0811 (1) 94.0656 (4) 82.0655 (1)
Lycopsamine isomer (isomer of **3**)	5.91	$[C_{15}H_{25}NO_5 + H]^+$	300.1805	300.1803	300.1803 (7) 210.1485 (1) 156.1019 (42) 139.0992 (14) 138.0914 (26) 120.0809 (18) 112.0759 (2) 108.0812 (2) 96.0813 (5) 95.0734 (6) 94.0656 (100) 82.0657 (5)
* Helioamplexine isomer I (isomer of **8**)	8.51	$[C_{16}H_{27}NO_5 + H]^+$	314.1962	314.1958	314.1958 (7) 270.1704 (0.5) 224.1643 (2) 156.1019 (44) 139.0992 (19) 138.0914 (25) 120.0810 (18) 112.0759 (2) 96.0812 (5) 95.0734 (8) 94.0656 (100) 82.0657 (7)
* Helioamplexine isomer II (isomer of **8**)	9.30	$[C_{16}H_{27}NO_5 + H]^+$	314.1962	314.1958	314.1960 (5) 270.1693 (0.5) 224.1643 (2) 156.1019 (33) 139.0992 (14) 138.0914 (25) 120.0809 (18) 112.0759 (2) 96.0812 (4) 95.0733 (7) 94.0656 (100) 82.0657 (6)
* Helioamplexine isomer (C7-isomer of **8**)	8.34	$[C_{16}H_{27}NO_5 + H]^+$	314.1962	314.1959	314.1959 (4) 156.1019 (100) 139.0992 (2) 138.0914 (6) 126.0916 (1) 120.0808 (2) 112.0759 (5) 108.0810 (10) 94.0655 (1) 86.0605 (2) 82.0656 (1) 80.0500 (1)
* 3′-*O*-Acetylintermedine	8.91	$[C_{17}H_{27}NO_6 + H]^+$	342.1911	342.1924	342.1924 (1) 282.1700 (4) 156.1019 (16) 139.0991 (7) 138.0913 (17) 120.0809 (20) 108.0812 (1) 96.0812 (2) 96.0450 (2) 95.0734 (3) 94.0655 (100) 82.0657 (3)
* 3′-*O*-Acetyllycopsamine	9.64	$[C_{17}H_{27}NO_6 + H]^+$	342.1911	342.1905	342.1905 (1) 282.1703 (2) 156.1019 (14) 139.0992 (3) 138.0913 (19) 120.0809 (19) 108.0812 (1) 96.0812 (2) 96.0448 (1) 95.0734 (2) 94.0655 (100) 82.0657 (3) 67.0549 (1)
Ideamine A (**10**) isomer I	2.43	$[C_{14}H_{23}NO_5 + H]^+$	286.1649	286.1648	286.1648 (11) 242.1392 (1) 196.1335 (2) 156.1019 (43) 139.0992 (11) 138.0914 (26) 120.0809 (18) 112.0758 (2) 108.0810 (2) 96.0813 (7) 95.0734 (5) 94.0656 (100) 82.0657 (5)
Ideamine A (**10**) isomer II	3.70	$[C_{14}H_{23}NO_5 + H]^+$	286.1649	286.1649	286.1649 (8) 156.1019 (35) 139.0992 (8) 138.0913 (30) 120.0808 (18) 96.0813 (4) 95.0734 (4) 94.0655 (100) 94.0593 (1) 82.0657 (3) 80.0500 (1) 67.0547 (1)
Ideamine A (**10**) isomer III	4.37	$[C_{14}H_{23}NO_5 + H]^+$	286.1649	286.1649	286.1649 (0) 156.1018 (45) 139.0992 (7) 138.0913 (26) 120.0807 (17) 98.2419 (3) 95.0733 (3) 94.0656 (100) 93.0705 (3) 82.0655 (4) 67.0549 (4)
Ideamine A (**10**) isomer IV	5.10	$[C_{14}H_{23}NO_5 + H]^+$	286.1649	286.1651	286.1651 (16) 258.9584 (3) 190.8082 (4) 188.1787 (3) 158.5305 (3) 156.1016 (48) 138.0913 (29) 124.0405 (3) 120.0807 (18) 94.0655 (100)
Ideamine A C7 isomer	1.72	$[C_{14}H_{23}NO_5 + H]^+$	286.1649	286.1656	286.1656 (10) 156.1019 (100) 138.0915 (5) 124.0757 (2) 120.0813 (1) 112.0760 (5) 108 0811 (7) 106.0656 (4) 94.0651 (2) 86.0603 (1)
Lycopsamine *N*-oxide isomer	7.61	$[C_{15}H_{25}NO_6 + H]^+$	316.1755	316.1751	316.1754 (46) 172.0967 (100) 155.0940 (21) 154.0862 (13) 138.0914 (94) 137.0836 (8) 136.0757 (25) 112.0759 (8) 111.0681 (14) 108.0811 (6) 94.0656 (35) 93.0577 (24)

Table 4. *Cont.*

Alkaloid	Typical R_T (min)	Molecular Ion Formula	Calculated $[M+H]^+$	Observed $[M+H]^+$	Mass Spectral Data (Rel. Int. %
7-Isomer of lycopsamine *N*-oxide	7.19	$[C_{15}H_{25}NO_6 + H]^+$	316.1755	316.1751	316.1751 (5) 172.0968 (100) 155.0940 (6) 154.0863 (2) 138.091 (3) 137.0836 (3) 136.0757 (2) 111.0681 (11) 106.0654 (6) 102.055 (2) 94.0657 (1)
Helioamplexine *N*-oxide isomer I	9.87	$[C_{16}H_{27}NO_6 + H]^+$	330.1911	330.1908	330.1908 (39) 286.1642 (4) 240.159 (7) 172.0967 (100) 155.0940 (30) 154.0862 (11) 138.0914 (84) 136.0758 (22) 112.0760 (8) 111.068 (18) 94.0655 (36) 93.0577 (20) 82.0419 (9)
Helioamplexine *N*-oxide isomer II	10.63	$[C_{16}H_{27}NO_6 + H]^+$	330.1911	330.1909	330.1909 (46) 172.0967 (100) 155.0940 (30) 154.0862 (12) 138.0914 (91) 137.0835 (9) 136.075 (23) 112.0759 (8) 111.0681 (20) 94.0655 (38) 93.0577 (26) 82.0419 (11)
3′-*O*-Acetyllycopsamine *N*-oxide isomer I	10.25	$[C_{17}H_{27}NO_7 + H]^+$	358.1860	358.1856	358.1856 (15) 316.1758 (4) 298.164 (40) 172.0968 (100) 155.0941 (23) 154.0863 (10) 138.0914 (73) 136.0758 (21) 111.0682 (19) 94.065 (33) 93.0577 (21) 89.0602 (19) 87.0446 (14)
3′-*O*-Acetyllycopsamine *N*-oxide isomer II	10.92	$[C_{17}H_{27}NO_7 + H]^+$	358.1860	358.1859	358.1859 (19) 316.1748 (4) 298.164 (39) 172.0967 (100) 155.0940 (19) 154.0862 (14) 138.0914 (89) 137.0835 (8) 136.0758 (27) 112.075 (7) 111.0681 (16) 94.0656 (44) 93.0577 (31)
7-*O*-Acetyllycopsamine *N*-oxide	10.03	$[C_{17}H_{27}NO_7 + H]^+$	358.1860	358.1857	358.1857 (21) 214.1073 (100) 180.1019 (48) 178.0863 (14) 137.0836 (53) 136.0758 (17) 120.0810 (16) 119.0731 (14) 106.0654 (13) 101.0601 (25) 89.0602 (15) 73.0291 (20)
Ideamine A *N*-oxide isomer I	4.87	$[C_{14}H_{23}NO_6 + H]^+$	302.1598	302.1595	302.1595 (67) 212.1275 (4) 172.0966 (91) 155.0939 (18) 154.0864 (10) 138.0913 (100) 137.0834 (7) 136.0757 (20) 112.0759 (7) 111.0681 (11) 108.0810 (6) 94.0655 (31) 93.0577 (23)
Ideamine A *N*-oxide isomer II	6.27	$[C_{14}H_{23}NO_6 + H]^+$	302.1598	302.1596	302.1596 (65) 172.0968 (96) 158.1181 (7) 155.0941 (18) 154.0861 (12) 138.0913 (100) 137.0835 (10) 136.0757 (26) 112.0759 (9) 111.0680 (9) 94.0656 (31) 93.0577 (24)
* 3′-*O*-Glucosyllycopsamine	8.94	$[C_{21}H_{35}NO_{10} + H]^+$	462.2336	462.2336	300.1806 (12) 156.1021 (16) 139.0994 (4) 138.0915 (40) 120.0810 (16) 97.0289 (3) 96.0813 (2) 96.0451 (2) 94.0657 (100) 85.0290 (7)
* 3′-*O*-Glucosylintermedine	7.41	$[C_{21}H_{35}NO_{10} + H]^+$	462.2336	462.2335	300.1807 (17) 156.1019 (20) 139.0993 (5) 138.0914 (48) 120.0810 (19) 97.0289 (4) 96.0813 (2) 94.0656 (100) 91.0580 (2) 85.0290 (10)
* 3′-*O*-Glucosyllycopsamine *N*-oxide	9.89	$[C_{21}H_{35}NO_{11} + H]^+$	478.2283	478.2286	316.1754 (100) 172.0968 (90) 155.0941 (15) 154.0860 (10) 138.0914 (74) 137.0837 (6) 136.0758 (19) 120.0809 (6) 112.0762 (5) 111.0680 (7) 94.0657 (47) 93.0578 (12)
* 3′-*O*-Glucosylintermedine *N*-oxide	8.58	$[C_{21}H_{35}NO_{11} + H]^+$	478.2283	478.2289	316.1758 (100) 172.0968 (76) 155.0943 (13) 154.0863 (9) 138.0916 (56) 137.0838 (7) 136.0757 (15) 120.0808 (7) 111.0681 (7) 94.0656 (29) 93.0578 (12)

Interestingly, PAs tentatively assigned as 3′-*O*-glucosyllycopsamine (found 462.2336, calculated for $C_{21}H_{35}NO_{10}+H^+$: 462.2336) and 3′-*O*-glucosylintermedine (found 462.2335, calculated for $C_{21}H_{35}NO_{10}+H^+$: 462.2336) and the corresponding *N*-oxides (found 478.2286 and 478.2289, calculated for $C_{21}H_{35}NO_{11}+H^+$: 478.2283), were also identified in minor amounts in the pods and nectar

(Table 4). The MS^2 spectra exhibited virtually identical MS^2 to the parent alkaloids **1** and **3**, **4** and **6**. A 3′-glucopyranosyl 2,3-dihydro-1*H*-pyrrolizin-1-one derivative has previously been reported from *Cynoglossum gansuense* [57]. Additionally, five isomeric components with MH^+ 286.1649 were also detected in *P. straminea* reduced extracts, with MS data consistent with these being desmethyl analogues of lycopsamine, i.e ideamine A (**10**) isomers (four esterified at C9, one at C7). Ideamine A *N*-oxide has previously been found in insects feeding on *Parsonsia laevigata* leaves [58,59]. Tessellatine (**9**), 3′-*O*-acetyl- and 7-*O*-acetyllycopsamine/intermedine and their *N*-oxides have been previously identified in *Amsinckia* or *Cryptantha* species [54,55,60,61]. To positively identify the PAs in lycopsamine-rich honey samples as originating from *Parsonsia straminea*, we sought to find some of these same minor PA components of this plant in honey.

2.8. Honey PA Profiles Linked to P. straminea

The detection of minor alkaloids in *Parsonsia straminea* provides a distinctive PA fingerprint in its HRAM LC-MS/MS profile, albeit in minor quantities compared to the major alkaloid lycopsamine (**3**). By comparison with the PA profile observed in market honey samples, there is clear evidence that this plant species is being used as a honey floral source by bees (Figure 7).

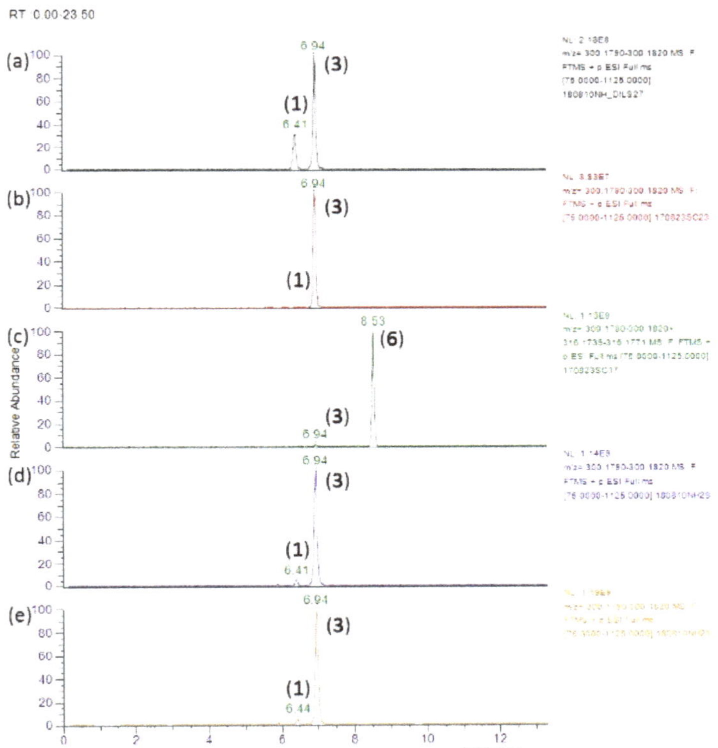

Figure 7. HRAM LC-MS/MS chromatograms (*m/z* 300.1805, *m/z* 316.1755) comparing the major pyrrolizidine alkaloids in *Parsonsia straminea* and honey: (**a**) intermedine (**1**) and lycopsamine (**3**) in *Parsonsia straminea* leaves (Zn reduced) (**b**) intermedine (**1**) and lycopsamine (**3**) *in Parsonsia straminea* flowers (Zn reduced) (**c**) lycopsamine (**3**) and its N-oxide (**6**) in *Parsonsia straminea* flowers (unreduced) (**d**) intermedine (**1**) and lycopsamine (**3**) in honey sample H-PA#146 (**e**) intermedine (**1**) and lycopsamine (**3**) in honey sample H-PA#157.

Honey samples such as H-PA#146 and H-PA#157 were independently purchased. When these honey samples were analysed against the 30 PA standards in our screen (Table 1), only the major alkaloid lycopsamine (**3**) and lesser intermedine (**1**) (and their *N*-oxides) were detected (Table 3). Characteristic major/minor components present in certain honeys (Table 4) included in addition to lycospamine and intermedine, the helioamplexine isomers at RT 8.20, 8.44 and 9.38 min (Figure 8) and putative 3′-*O*-acetylintermedine (8.91 min) and 3′-*O*-acetyllycospamine (9.64 min). The tentatively assigned 3′-*O*-glucosylintermedine and 3′-*O*-glucosyllycopsamine were also identified in these honey samples with the MS^2 spectra observed identical to that found in the plant pods and nectar. Non-toxic dihydrolycopsamine isomers were also identified in the plant and honey. Due to the low levels of these minor PAs in the plant, they were seen most readily in honey samples highest in lycopsamine (**3**) (eg., H-PA#19,157,146).

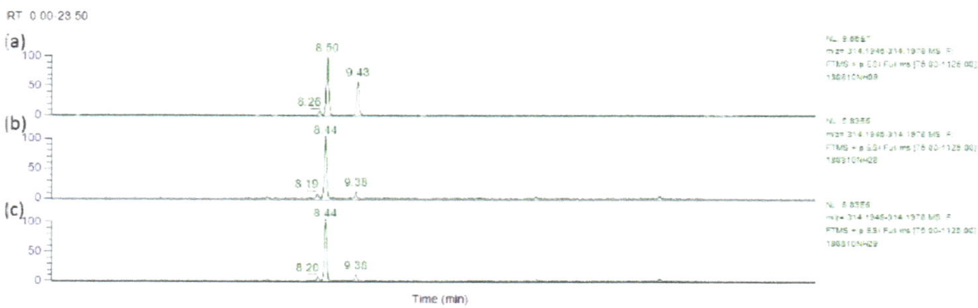

Figure 8. HRAM LC-MS/MS chromatograms (*m/z* 314.1911) comparing minor peaks in *Parsonsia straminea* and honey (**a**) isomers of helioamplexine (**8**) in *Parsonsia straminea* leaves (Zn reduced) (**b**) isomers of (**8**) in honey sample H-PA#146 (**c**) isomers of (**8**) in honey sample H-PA#157.

2.9. Plant Origins of PAs in Honeys Surveyed

Of the 30 PA standards utilised in our survey, fifteen PAs (50%) were not detected in any of the market honey samples (Table 5). As shown in Table 5, based on profiles of alkaloids identified most of the honey PAs were likely sourced from *Parsonsia straminea* or *Heliotropium amplexicaule*, with honey containing *Parsonsia* alkaloids being dominant in lycopsamine (**3**) (up to 3100 ng/g) and honey containing *Heliotropium amplexicaule* alkaloids dominant in indicine (**2**) (up to 1700 ng/g). PAs sourced from *Echium plantagineum* were much lower, with the dominant PA detected being echimidine (**7**) (up to 260 ng/g) in agreement with previous studies [11,62]. Even lower levels of PAs from *Heliotropium europaeum* (containing lasiocarpine, heliotrine and europine [11,62] (and their *N*-oxides)) and Senecio species (most likely *Senecio madagascariensis*) [63] were detected (Table 5). Of course, many of the honey samples are ascribed by their label to particular non-PA producing floral sources, so the observation of PAs in these honeys is a product of either the natural foraging of bees on different available plants, or the blending of honeys in the packaging process. This co-foraging/blending is also evident in honey samples that show co-occurrence of pyrrolizidine alkaloids from multiple floral sources, for example, honeys containing indicine (**2**) (from *H. amplexicaule*) and lycopsamine (**3**) (likely from *P. straminea* due to lack of the dominant PA echimidine (**7**) as present in *E. plantagineum*). Both these sets of PAs were present in significant levels in H-PA#11, 32, 216, 630 and 642. Geographically both the low-growing heliotrope, *H. amplexicaule*, and the arboreal vine, *P. straminea*, can co-occur in sub-tropical coastal regions of Queensland [53,64], so the co-occurrence of their respective alkaloids in honey would seem logical if both plants are visited by foraging bees within the same landscape. The high abundance of alkaloids from these quite different plant species in honey suggests that both are attractive to foraging bees, and where possible both species should be avoided when siting honey hives. It is apparent that the 'standard set' for PA/PANO testing of honeys varies depending on the natural flora of the region, as well as the cultivated plants present. In this study erucifoline, jacobine, monocrotaline, senciphylline, or their corresponding *N*-oxides and senkirkine or trichodesmine were not found in the honey tested, which is a considerably different result to those found recently in Schleswig-Holstein region of Germany [28].

Table 5. Number of PA positive samples and PA concentration in honey samples (n = 465), grouped by potential source of PA plant origin (mean and median are for positive samples only).

Potential Plant Source of PAs in Honey	Pyrrolizidine Alkaloids	Number of Honey Samples	PA Content (ng/g)		
			Mean	Median	Max
Parsonsia straminea (Monkey rope)					
	Lycopsamine	274	210	33	5100
	Lycopsamine *N*-oxide	31	320	51	1900
	Intermedine	217 [b]	32	18	290
	Intermedine *N*-oxide [a]	30 [c]	86	21	560
Heliotropium amplexicaule (Blue heliotrope)					
	Indicine	221	120	31	1700
	Indicine *N*-oxide/intermedine *N*-oxide (n.r.) [a]	30 [c]	86	21	660
	Intermedine	217 [b]	32	18	290
	Helioamplexine [d]	78	23	13	130
	Helioamplexine *N*-oxide [e]	5	28	35	51
Echium plantagineum (Paterson's curse)					
	Echimidine	93	36	20	260
	Lycopsamine	76 [f]			
	Intermedine	37 [g]			
Heliotropium europaeum (Potato weed)					
	Europine	26	30	11	160
	Europine *N*-oxide	4	11	7	24
	Heliotrine	17	23	15	75
	Heliotrine *N*-oxide	1	9	9	9
	Lasiocarpine	10	20	14	59
***Senecio* species**					
	Senecivernine	19	24	14	150
	Senecionine	7	13	8	41
	Retrorsine	18	20	12	91
No occurrence/Below LOR					
	Echimidine *N*-oxide	0			
	Erucifoline	0			
	Erucifoline *N*-oxide	0			
	Jacobine	0			
	Jacobine *N*-oxide	0			
	Lasiocarpine *N*-oxide	0			
	Monocrotaline	0			
	Monocrotaline *N*-oxide	0			
	Retrorsine *N*-oxide	0			
	Senecionine *N*-oxide	0			
	Seneciphylline	0			
	Seneciphylline *N*-oxide	0			
	Senecivernine *N*-oxide	0			
	Senkirkine	0			
	Trichodesmine	0			

[a] Not resolved–indicine *N*-oxide and intermedine *N*-oxide co-eluted. [b,c] Intermedine and intermedine *N*-oxide are present in multiple plants, and prominent in both *Parsonsia straminea* and *Heliotropium amplexicaule*. [d] Helioamplexine was quantified using heliotrine standard curve. [e] Helioamplexine *N*-oxide was quantified using heliotrine *N*-oxide standard curve. [f] Lycopsamine observed in honey containing echimidine not necessarily attributed solely to *E. plantagineum*, but of the 93 honey samples containing echimidine, the concentration of lycopsamine was lower than echimidine in 76 honeys, which is consistent with the relative amounts observed in *E. plantagineum*. In the other 17 honeys which contain echimidine, it is likely that there is more than one source of lycopsamine. [g] Of the 76 honeys in which lycopsamine was at a lower level than echimidine, 37 honey samples also contained intermedine.

2.10. *Honey as a Dietary Source of Pyrrolizidine Alkaloids*

Major supermarket honeys by comparison represent blended honeys from diverse locations, some of which attributed the specific floral source and in general contained only low levels of PAs. It has been observed previously that blended retail honeys had a lower PA content, but that PAs were present in more samples [65]. In this study, for supermarket honeys (n = 129), PAs were detected in 84% of honeys, and showed highest total PA levels of 1400 ng/g. For supermarket honeys, the mean total PA level of PA-positive samples was 120 ng/g and the median level was 61 ng/g.

Certain small producer honeys displayed the highest levels of pyrrolizidine alkaloids, with the PA content dependent on the location and attractiveness of PA containing plants to foraging honey bees. Paradoxically, even though analysed PA content of small producer honeys range from <LOR to an alarming 3000 ng/g, if equal amounts of each of these 205 small producer honeys were blended, the hypothetical resultant mixed honey would have a PA content of only 240 ng/g (i.e., the average PA content of all of these 465 honeys).

It has been observed previously in South American honeys that raw honeys showed greater variety due to the availability of PA containing plants near to hives [65].

The cumulative toxicity of the 1,2-unsaturated PAs have been demonstrated in animal studies and genotoxicities/tumorigenicities were induced by hepatic metabolism of PAs [66]. Consequently, provisional tolerable daily intakes (PTDI) have been recommended to control the human consumption of PAs [5,10,39,67].

Using the Australian FSANZ provisional tolerable daily intake (PTDI) of 1 μg/kg BW/day, 0% of honeys tested (total n = 465) exceeded the limit for a 70 kg adult consuming 20 g of honey per day, but 19% of honeys tested exceeded the limit for a 15 kg child consuming 50 g of honey per day. Applying the lowest recommended PTDI (EFSA, COT, BfR) of 0.007 μg/kg BW/day, 63% of honeys tested exceeded the limit for a 70 kg adult consuming 20 g of honey per day and 84% of honeys tested exceeded the limit for a 15 kg child consuming 50 g of honey per day.

The PA content of honey samples varies with geographical location and climate, determined by the type and distribution of PA containing plants and by the propensity for bees to forage on these plants [65,68]. Lycopsamine (**3**) and intermedine (**1**) are present in many PA-producing plants, with the knowledge of the plants distributed in Australia and the ratio to other PAs present, it is likely that *Parsonsia straminea* is a major contributor to the high PA levels observed in certain honeys in this study. Of course, it is possible that there is more than one PA source of lycopsamine (**3**), with a small portion of lycopsamine contamination of honey potentially originating from *Echium plantagineum* and *Ageratum houstonianium*. Also, there are likely other PA containing plants that have not been considered. It is also possible that not all PAs present in honey have been identified by comparison with standards and by analysis of the top MSMS. Despite the observation that of the PAs tested in experimental rats, lycopsamine (**3**) induced the lowest levels of liver DNA adducts (formed from PA derived reactive pyrrolic metabolites), PA containing plants are the most common poisonous/carcinogenic plants affecting livestock, wildlife and humans [69]. Beekeepers are advised to avoid these known plant genera around the hive/apiary as much as possible to reduce PA contamination in honey.

3. Conclusions

The HRAM LC-MS/MS method for pyrrolizidine alkaloid analysis described here enables the ready resolution of isomeric alkaloids of the lycopsamine-type. The described simple adjustment of column conditions to a lower temperature was effective in resolving the problematic pairs of indicine/lycopsamine alkaloids present in Australian honey. This resolution has enabled us to identify *Parsonsia* vines as a previously unsuspected source of PA contamination in Australian honey. Low temperature chromatographic resolution may have as yet unexplored application in resolving other similar diastereomeric pyrrolizidine alkaloid isomers, of which there are many within the known pyrrolizidine alkaloids, many of which do not have commercially available standards.

4. Materials and Methods

4.1. Chemicals and Solvents

In total, 30 pyrrolizidine alkaloid standards were utilized in a high resolution accurate mass (HRAM) LC-MS/MS screen. Echimidine, erucifoline, europine, heliotrine, indicine, intermedine, jacobine, lasiocarpine lycopsamine, monocrotaline, retrorsine senecionine, seneciphylline, senecivernine, and their respective *N*-oxides, were purchased together with senkirkine and trichodesmine from Phytolab GmbH & Co. KG (Vestenbergsgreuth, Germany) and had a purity >89%. All other chemicals and solvents were of analytical reagent or HPLC grade purity. Water used for sample preparation and HPLC was Milli-Q purified (Merck Millipore, Darmstadt, Germany).

4.2. Honey Samples

Honey samples (465 in total) were purchased between September 2016 and December 2017 directly from Queensland supermarkets, fruit shops, local markets, and producers.

4.3. Honey Alkaloid Extraction

Honey samples (1 g) were dissolved in aqueous H_2SO_4 (0.05 M, 10 mL) centrifuged and the supernatant applied to preconditioned Agilent SPE Bond Elut 100 mg LRC-SCX columns (Agilent Technologies, Folsom, CA, USA). SPE cartridges were washed with water (10 mL) and methanol (10 mL), and pyrrolizidine alkaloids were then eluted with 3% ammonia in methanol (3 mL). The eluate was evaporated to dryness under nitrogen, and the residue reconstituted in 5% methanol in water (1 mL) for HRAM LC-MS/MS analysis.

4.4. Honey Method Validation

The validation of the method was conducted according to the National Association of Testing Authorities (NATA) guidance document [22]. The method was validated in 3 blank honeys, and based on results for 10 spiked samples at a spiking level of 5 ng/g (Table 2), 10 blank samples and 10 non-extracted spike samples and the recoveries determined. Limit of detection (LOD) was calculated as 3*s*. Limit of quantitation was calculated as 9*s*. Limit of reporting was set at the levels the samples were spiked, also the level of the lowest standard used for the calibration curve. The uncertainties given are at the 95% confidence level as required by the NATA [22,23]. Replicate samples were prepared for every tenth honey test sample to assess reproducibility. The difference between replicate samples (coefficient of variance %) was typically 0.12–6.7%. High samples were diluted to levels within the calibration curve and re-run. SPE wash steps and further elutions with 3% ammonia in methanol (3 mL) were analysed for residual PAs and the extraction was found to be exhaustive. Table 2 shows good recoveries for most PAs.

4.5. Plant Alkaloid Extraction

4.5.1. Plant Source

Parsonsia straminea was collected from a suburban area in the south of Brisbane and was taxonomically identified by the Queensland Herbarium, with a voucher specimen (AQ522465) incorporated into their collection. The *Parsonsia straminea* foliage sample was a collection of stems and leaves, and was freeze dried, milled and stored frozen prior to analysis. Pods were collected separately and freeze-dried, milled and frozen. Flowers were sampled as both intact flowers (freeze-dried, milled and frozen) or utilized to provide nectar and pollen separately. Nectar was separated from flowers using a microcap capillary, and anthers and pollen were separated from other flower plants with tweezers and desiccated.

4.5.2. Foliage and Seed Pod Extracts

Dried milled plant leaves and stems (1 g) and seed pods (1 g) were separately dissolved in methanol (10 mL), vortexed (20 s), shaken (30 min) then centrifuged (4800 rpm, 10 min) and the supernatants removed and concentrated to dryness under nitrogen. The residues were dissolved in aqueous H_2SO_4 (0.05 M, 10 mL) centrifuged (4800 rpm, 10 min) and a portion of the supernatants (0.1 mL) were applied to preconditioned Agilent SPE Bond Elut 500 mg LRC-SCX columns. SPE cartridges were washed with water (10 mL) and methanol (10 mL), and pyrrolizidine alkaloids were then eluted with 3% ammonia in methanol (10 mL). The eluate was evaporated to dryness under nitrogen, and the residue reconstituted in 5% methanol in water (1 mL) for HRAM LC-MS/MS analysis.

4.5.3. Whole Flower Extracts

Dried and milled flowers (0.1 g) were dissolved in methanol (2 mL), vortexed (20 s), shaken (30 min) then centrifuged (4800 rpm, 10 min) and the supernatant removed and concentrated to dryness under nitrogen. The residue was dissolved in aqueous H_2SO_4 (0.05 M, 1 mL), centrifuged (4800 rpm, 10 min) and, for each sample, a portion of the supernatant (0.1 mL) was applied to a preconditioned Agilent SPE Bond Elut 500 mg LRC-SCX column. Each SPE cartridge was washed with water (10 mL) and methanol (10 mL), and pyrrolizidine alkaloids were then eluted with 3% ammonia in methanol (10 mL). The eluate was evaporated to dryness under nitrogen, and the residue reconstituted in 5% methanol in water (1 mL) for HRAM LC-MS/MS analysis.

4.5.4. Zinc Reduced Extracts

Another portion of each aqueous H_2SO_4 supernatant (0.5 mL) derived from leaves/stems, pods and flowers was treated with Zn dust (100 mg) and stirred (2 h). After centrifugation (4800 rpm, 10 min), a portion of the supernatants (0.1 mL) was applied to a preconditioned Agilent SPE Bond Elut 500 mg LRC-SCX column. Each SPE cartridge was washed with water (10 mL) and methanol (10 mL), and pyrrolizidine alkaloids were then eluted with 3% ammonia in methanol (10 mL). The eluate was evaporated to dryness under nitrogen, and the residue reconstituted in 5% methanol in water (1 mL) for HRAM LC-MS/MS analysis.

4.5.5. Floral Nectar Extract

Nectar (22.5 mg) was obtained from fresh flowers using a microcap capillary, dissolved in MeOH (0.5 mL) and diluted 1 in 100 with 5% methanol in water (1 mL) for HRAM LC-MS/MS analysis.

4.5.6. Pollen Extract

Dessicated anthers and pollen were placed in hexane and shaken (1 min). The hexane containing pollen was separated from the anthers and evaporated to dryness under nitrogen. The resulting pollen (0.36 mg) was dissolved in 5% methanol in water and diluted as required for LCMS/MS analysis.

4.6. HRAM LC-MS/MS Analysis

Samples were analysed using a Vanquish UHPLC in combination with Q Exactive Orbitrap high resolution accurate mass (HRAM) spectrometry system (Thermo Fisher Scientific, Bremen, Germany). LC-MS/MS separation was achieved on a Kinetex XB-C18 analytical column (100 × 2.1 mm, 2.6 μm, 100 Å) at 5 °C. Analysis conditions: binary solvent system, solvent A (ammonium formate (5 mM) and formic acid (0.1%) and solvent B (95% *v*/*v* methanol/water with ammonium formate (5 mM) and formic acid (0.1%)). Compounds were eluted from the column at 0.3 mL min^{-1} with mobile phase B held at 5% for 3 min followed by linear gradients of B from 5–50% (3–15 min), 50–80% (15–18.5 min), 80–100% (18.5–19 min), where it was held for 30 sec, before reducing from 100–5% over 6 sec, where it was held until stop at 23.5 min. Instrument control, data acquisition and analysis were conducted using Tracefinder 4.1 from Thermo Fisher Scientific. Alkaloid detection was performed by positive electrospray ionisation (ESI) with a spray voltage of 3500 V and a vaporiser temperature of 400 °C. MS analysis run with arbitrary pressures of sheath gas 48, aux gas 11, sweep gas 2, spray voltage 3.5 kV, capillary temperature of 320 °C, auxiliary gas heater at 350 °C and used full scan/dd-MS^2 mode. Full scans were conducted at a resolution of 70,000 FWHM (at *m*/*z* 200), with an AGC target of 1 00 × 10^6. The maximum time of accumulating ions per scan event was 10 ms with a scan range of 75–1125 *m*/*z*. Data dependent acquisition (dd-MS^2) was conducted at a resolution of 17,500 with an AGC target of 1.00 × 10^6. The maximum time of accumulating ions per scan event was 50 ms. Normalized collision energy (nce) was set to 50% and an isolation window of 1.0 *m*/*z* was utilized. Dynamic exclusion was set to 3 s preventing subsequent triggering of the same ion in data dependent scans. A maximum of 5 most abundant precursors could be selected for dd-MS^2 per scan event.

Pyrrolizidine alkaloid levels in honey/plant material were quantitated against certified PA standards, with calibration curves obtained for each of the 30 pyrrolizidine alkaloid standards injected at 5, 10, 20, 50 and 100 and 200 ppb (in duplicate/triplicate). Honey or plant extracts were analysed by HRAM LC-MS/MS to detect pyrrolizidine alkaloids and their *N*-oxides by matching of retention time with the corresponding standard and identified by their precursor parent ion (M+H$^+$) and confirmed by the detection of product ions (Table 1). The identity of these and further alkaloids was assigned by use of the high resolution accurate mass data provided by the Q Exactive mass spectrometer, enabling the determination of elemental composition of parent and fragment ions (Tables 3 and 4).

Author Contributions: Concept, M.T.F.; methodology, N.L.H.; validation, B.L.L.T., D.H., K.J.M.; formal analysis, S.J.C., S.R.A.; investigation, N.L.H.; extraction N.L.H., C.L.M., E.S., M.Y., T.T.P.N.; writing—original draft preparation, N.L.H., M.T.F.; writing—review and editing, N.L.H., M.T.F., K.J.M.; supervision, M.T.F.; funding acquisition, M.T.F., K.J.M.

Funding: This work was funded by Queensland Health Grant RSS17-002.

Conflicts of Interest: The authors declare no conflict of interest.

References

1. Avula, B.; Sagi, S.; Wang, Y.-H.; Zweigenbaum, J.; Wang, M.; Khan, I.A. Characterization and screening of pyrrolizidine alkaloids and N-oxides from botanicals and dietary supplements using UHPLC-high resolution mass spectrometry. *Food Chem.* **2015**, *178*, 136–148. [CrossRef]
2. Smith, L.W.; Culvenor, C.C.J. Plant sources of hepatotoxic pyrrolizidine alkaloids. *J. Nat. Prod.* **1981**, *44*, 129–152. [CrossRef]
3. Stegemann, T.; Kruse, L.H.; Brutt, M.; Ober, D. Specific Distribution of Pyrrolizidine Alkaloids in Floral Parts of Comfrey (*Symphytum officinale*) and its Implications for Flower Ecology. *J. Chem. Ecol.* **2019**, *45*, 128–135. [CrossRef]
4. Edgar, J.A.; Colegate, S.M.; Boppre, M.; Molyneux, R.J. Pyrrolizidine alkaloids in food: A spectrum of potential health consequences. *Food Addit. Contam. Part. A* **2011**, *28*, 308–324. [CrossRef]
5. FSANZ. *Pyrrolizidine Alkaloids in Food. A Toxicological Reveiw and Risk Assessment*; Technical Report Series No. 2; Australia New Zealand Food Authority: Canberra, Australia, 2001. Available online: www.foodstandards.gov.au/publications/documents/TR2.pdf (accessed on 17 December 2019).
6. Alexander, J.; Benford, D.; Boobis, A.; Ceccatelli, S.; Cottrill, B.; Cravedi, J.-P.; Di Domenico, A.; Doerge, D.; Dogliotti, E.; Edler, L.; et al. Scientific opinion on pyrrolizidine alkaloids in food and feed. *EFSA J.* **2011**, *9*, 2406. [CrossRef]
7. COT. COT Statement on Pyrrolizidine Alkaloids in Food. Available online: https://cot.food.gov.uk/sites/default/files/cot/cotstatementpa200806.pdf (accessed on 2 August 2019).
8. BfR. Bundesinstitut für Risikobewertung (The Federal Instituteof Risk Assessment). International Collaborative Study for the Determination of Pyrrolizidine Alkaloids in Honey and Herbal Tea by SPE-LC-MS/MS. Available online: http://www.bfr.bund.de/cm/350/international-collaborative-study-for-the-determination-of-pyrrolizidine-alkaloids-in-honey-and-herbal-tea-by-spe-lc-ms-ms.pdf (accessed on 19 April 2016).

9. Dusemund, B.; Nowak, N.; Sommerfeld, C.; Lindtner, O.; Schäfer, B.; Lampen, A. Risk assessment of pyrrolizidine alkaloids in food of plant and animal origin. *Food Chem. Toxicol.* **2018**, *115*, 63–72. [CrossRef] [PubMed]
10. European Food Safety Authority. Risks for human health related to the presence of pyrrolizidine alkaloids in honey, tea, herbal infusions and food supplements. *EFSA J.* **2017**, *15*, 4908. [CrossRef]
11. Beales, K.A.; Betteridge, K.; Colegate, S.M.; Edgar, J.A. Solid-phase extraction and LC-MS analysis of pyrrolizidine alkaloids in honeys. *J. Agric. Food. Chem.* **2004**, *52*, 6664–6672. [CrossRef] [PubMed]
12. Griffin, C.T.; Mitrovic, S.M.; Danaher, M.; Furey, A. Development of a fast isocratic LC-MS/MS method for the high-throughput analysis of pyrrolizidine alkaloids in Australian honey. *Food Addit. Contam. Part. A* **2014**, *32*, 214–228. [CrossRef] [PubMed]
13. Culvenor, C.C.J.; Edgar, J.A.; Smith, L.W. Pyrrolizidine alkaloids in honey from *Echium plantagineum* L. *J. Agric. Food Chem.* **1981**, *29*, 958–960. [CrossRef] [PubMed]
14. Julien, M.H. Biological control of rangeland weeds in Australia. *Rangel. J.* **2006**, *28*, 47. [CrossRef]
15. Fletcher, M.T.; McKenzie, R.A.; Blaney, B.J.; Reichmann, K.G. Pyrrolizidine alkaloids in *Crotalaria* taxa from northern Australia: Risk to grazing livestock. *J. Agric. Food. Chem.* **2009**, *57*, 311–319. [CrossRef] [PubMed]
16. Fletcher, M.T.; Hayes, P.Y.; Somerville, M.J.; De Voss, J.J. *Crotalaria medicaginea* associated with horse deaths in northern Australia: New pyrrolizidine alkaloids. *J. Agric. Food. Chem.* **2011**, *59*, 11888–11892. [CrossRef] [PubMed]
17. McKenzie, R.A. *Australia's Poisonous Plants, Fungi and Cyanobacteria*; CSIRO Publishing: Collingwood, Australia, 2012.
18. Edgar, J.A. Pyrrolizidine alkaloids sequestered by Solomon Island Danaine butterflies. The feeding preferences of the Danainae and Ithomiinae. *J. Zool.* **1982**, *196*, 385–399. [CrossRef]
19. Abe, F.; Nagao, T.; Okabe, H.; Yamauchi, T. Macrocyclic pyrrolizidine alkaloids from *Parsonsia laevigata*. *Phytochemistry* **1991**, *30*, 1737–1739. [CrossRef]
20. Honda, K.; Hayashi, N.; Abe, F.; Yamauchi, T. Pyrrolizidine Alkaloids Mediate Host–Plant Recognition by Ovipositing Females of an Old World Danaid Butterfly, Idea leuconoe. *J. Chem. Ecol.* **1997**, *23*, 1703–1713. [CrossRef]
21. Plants of the World Online (POWO). Facilitated by the Royal Botanic Gardens: Kew, UK. Available online: http://www.plantsoftheworldonline.org/ (accessed on 15 November 2019).
22. NATA General Accreditation Guidance—Validation and Verification of Quantitative and Qualitative Test Methods. January 2018. Available online: https://www.nata.com.au/phocadownload/gen-accreditation-guidance/Validation-and-Verification-of-Quantitative-and-Qualitative-Test-Methods.pdf (accessed on 22 May 2019).
23. NATA General Accreditation Guidance—Estimating and Reporting Measurement Uncertainty of Chemical Test Results. January 2018. Available online: https://www.nata.com.au/phocadownload/gen-accreditation-guidance/Estimating-and-reporting-measurement-uncertainty-of-chemical-test-results.pdf (accessed on 22 May 2019).
24. EU Reference Laboratories for Residues of Pesticides. Method Validation & Quality Control Procedures for Pesticide Residues Analysis in Food & Feed, SANTE/11813/2017. Available online: https://www.eurl-pesticides.eu/docs/public/tmplt_article.asp?CntID=727 (accessed on 20 November 2019).
25. Carpinelli de Jesus, M.; Hungerford, N.L.; Carter, S.J.; Anuj, S.R.; Blanchfield, J.T.; De Voss, J.J.; Fletcher, M.T. Pyrrolizidine Alkaloids of Blue Heliotrope (*Heliotropium amplexicaule*) and Their Presence in Australian Honey. *J. Agric. Food. Chem.* **2019**, *67*, 7995–8006. [CrossRef]
26. Wuilloud, J.C.; Gratze, S.R.; Gamble, B.M.; Wolnik, K.A. Simultaneous analysis of hepatotoxic pyrrolizidine alkaloids and N-oxides in comfrey root by LC-ion trap mass spectrometry. *Analyst* **2004**, *129*, 150–156. [CrossRef]
27. Kempf, M.; Wittig, M.; Reinhard, A.; von der Ohe, K.; Blacquiere, T.; Raezke, K.-P.; Michel, R.; Schreier, P.; Beuerle, T. Pyrrolizidine alkaloids in honey: Comparison of analytical methods. *Food Addit. Contam. Part. A* **2011**, *28*, 332–347. [CrossRef]
28. Gottschalk, C.; Huckauf, A.; Dubecke, A.; Kaltner, F.; Zimmermann, M.; Rahaus, I.; Beuerle, T. Uncertainties in the determination of pyrrolizidine alkaloid levels in naturally contaminated honeys and comparison of results obtained by different analytical approaches. *Food Addit. Contam. Part. A* **2018**, *35*, 1366–1383. [CrossRef]

29. Kaltner, F.; Stiglbauer, B.; Rychlik, M.; Gareis, M.; Gottschalk, C. Development of a sensitive analytical method for determining 44 pyrrolizidine alkaloids in teas and herbal teas via LC-ESI-MS/MS. *Anal. Bioanal. Chem.* **2019**. [CrossRef] [PubMed]
30. Wang, T.; Frandsen, H.L.; Christiansson, N.R.; Rosendal, S.E.; Pedersen, M.; Smedsgaard, J. Pyrrolizidine alkaloids in honey: Quantification with and without standards. *Food Control.* **2019**, *98*, 227–237. [CrossRef]
31. Tamošiūnas, V.; Mischke, C.; Mulder, P.P.J.; Stroka, J. EU Publications. In *Report on the 2012 Proficiency Test on Pyrrolizidine Alkaloids in Honey and Hay*; Institute for Reference Materials and Measurements (Joint Research Centre): Luxembourg, 2013. [CrossRef]
32. Mulder, P.P.J.; Sánchez, P.L.; These, A.; Preiss-Weigert, A.; Castellari, M. Occurrence of Pyrrolizidine Alkaloids in food. *EFSA Support. Publ.* **2015**, EN-859. [CrossRef]
33. Crews, C. Methods for Analysis of Pyrrolizidine Alkaloids. In *Natural Products*; Ramawat, K.G., Merillon, J.M., Eds.; Spinger: Berlin/Heidelberg, Germany, 2013; pp. 1049–1068.
34. Van de Schans, M.G.M.; Blokland, M.H.; Zoontjes, P.W.; Mulder, P.P.J.; Nielen, M.W.F. Multiple heart-cutting two dimensional liquid chromatography quadrupole time-of-flight mass spectrometry of pyrrolizidine alkaloids. *J. Chromatogr. A* **2017**, *1503*, 38–48. [CrossRef] [PubMed]
35. Griffin, C.T.; O'Mahony, J.; Danaher, M.; Furey, A. Liquid Chromatography Tandem Mass Spectrometry Detection of Targeted Pyrrolizidine Alkaloids in Honeys Purchased within Ireland. *Food Anal. Method.* **2015**, *8*, 18–31. [CrossRef]
36. Colegate, S.M.; Edgar, J.A.; Knill, A.M.; Lee, S.T. Solid-phase extraction and HPLC-MS profiling of pyrrolizidine alkaloids and their N-oxides: A case study of *Echium plantagineum*. *Phytochem. Anal.* **2005**, *16*, 108–119. [CrossRef] [PubMed]
37. Skoneczny, D.; Weston, P.A.; Zhu, X.; Gurr, G.M.; Callaway, R.M.; Weston, L.A. Metabolic Profiling of Pyrrolizidine Alkaloids in Foliage of Two *Echium* spp. Invaders in Australia—A Case of Novel Weapons? *Int. J. Mol. Sci.* **2015**, *16*, 26721–26737. [CrossRef]
38. Atlas of Living Australia website. *Echium plantagineum* L. Available online: https://bie.ala.org.au/species/http://id.biodiversity.org.au/node/apni/2889451. (accessed on 1 August 2019).
39. European Food Safety Authority. Scientific Opinion on Pyrrolizidine alkaloids in food and feed. *EFSA J.* **2011**, *9*, 2406. [CrossRef]
40. Atlas of Living Australia Website. Available online: https://www.ala.org.au (accessed on 1 August 2019).
41. Okunade, A.L. *Ageratum conyzoides* L. *Asteraceae Fitoterapia* **2002**, *73*, 1–16. [CrossRef]
42. Wiedenfeld, H.; Roder, E. Pyrrolizidine Alkaloids from *Ageratum conyzoides*. *Planta Med.* **1991**, *57*, 578–579. [CrossRef]
43. Molyneux, R.J.; Gardner, D.L.; Colegate, S.M.; Edgar, J.A. Pyrrolizidine alkaloid toxicity in livestock: A paradigm for human poisoning? *Food Addit. Contam. Part. A* **2011**, *28*, 293–307. [CrossRef] [PubMed]
44. Bosi, C.F.; Rosa, D.W.; Grougnet, R.; Lemonakis, N.; Halabalaki, M.; Skaltsounis, A.L.; Biavatti, M.W. Pyrrolizidine alkaloids in medicinal tea of *Ageratum conyzoides*. *Rev. Bras. Farmacogn.* **2013**, *23*, 425–432. [CrossRef]
45. Wiedenfeld, H.; Andrade-Cetto, A. Pyrrolizidine alkaloids from *Ageratum houstonianum* Mill. *Phytochemistry* **2001**, *57*, 1269–1271. [CrossRef]
46. Culvenor, C.C.J.; Smith, L.W. The alkaloids of Amsinckia species: *A. intermedia* Fisch. & Mey., *A. hispida* (Ruiz. & Pav.) Johnst. and *A. lycopsoides* Lehm. *Aust. J. Chem.* **1966**, *19*, 1955. [CrossRef]
47. Boppre, M.; Colegate, S.M. Recognition of pyrrolizidine alkaloid esters in the invasive aquatic plant *Gymnocoronis spilanthoides* (Asteraceae). *Phytochem. Anal.* **2015**, *26*, 215–225. [CrossRef]
48. Edgar, J.A.; Culvenor, C.C.J. Pyrrolizidine ester alkaloid in danaid butterflies. *Nature* **1974**, *248*, 614–616. [CrossRef]
49. Edgar, J.A.; Culvenor, C.C.J. Pyrrolizidine alkaloids in *Parsonsia* species (family Apocynaceae) which attract Danaid butterflies. *Experientia* **1975**, *31*, 393–394. [CrossRef]
50. Burzynski, E.A.; Minbiole, K.P.C.; Livshultz, T. New sources of lycopsamine-type pyrrolizidine alkaloids and their distribution in Apocynaceae. *Biochem. Syst. Ecol.* **2015**, *59*, 331–339. [CrossRef]
51. Orr, A.G.; Trigo, J.R.; Witte, L.; Hartmann, T. Sequestration of pyrrolizidine alkaloids by larvae of *Tellervo zoilus* (Lepidoptera: Ithomiinae) and their role in the chemical protection of adults against the spider *Nephila maculata* (Araneidae). *Chemoecology* **1996**, *7*, 68–73. [CrossRef]

52. Reinhard, A.; Janke, M.; Ohe, W.; Kempf, M.; Theuring, C.; Hartmann, T.; Schreier, P.; Beuerle, T. Feeding Deterrence and Detrimental Effects of Pyrrolizidine Alkaloids Fed to Honey Bees (*Apis mellifera*). *J. Chem. Ecol.* **2009**, *35*, 1086–1095. [CrossRef]
53. Atlas of Living Australia website. *Parsonsia straminea* (R.Br.) F. Muell. Available online: https://bie.ala.org.au/species/http://id.biodiversity.org.au/node/apni/2899618 (accessed on 21 February 2019).
54. Colegate, S.M.; Welsh, S.L.; Gardner, D.R.; Betz, J.M.; Panter, K.E. Profiling of dehydropyrrolizidine alkaloids and their N-oxides in herbarium-preserved specimens of *Amsinckia* species using HPLC-esi (+) MS. *J. Agric. Food. Chem.* **2014**, *62*, 7382–7392. [CrossRef] [PubMed]
55. Kelley, R.B.; Seiber, J.N. Pyrrolizidine alkaloid chemosystematics in *Amsinckia*. *Phytochemistry* **1992**, *31*, 2369–2387. [CrossRef]
56. These, A.; Bodi, D.; Ronczka, S.; Lahrssen-Wiederholt, M.; Preiss-Weigert, A. Structural screening by multiple reaction monitoring as a new approach for tandem mass spectrometry: Presented for the determination of pyrrolizidine alkaloids in plants. *Anal. Bioanal. Chem.* **2013**, *405*, 9375–9383. [CrossRef] [PubMed]
57. Jin, Y.-P.; Wei, X.-N.; Shi, Y.-P. Chemical Constituents from *Cynoglossum gansuense*. *Helv. Chim. Acta* **2007**, *90*, 776–782. [CrossRef]
58. Nishida, R.; Kim, C.-S.; Fukami, H.; Irik, R. Ideamine N-Oxides: Pyrrolizidine Alkaloids Sequestered by the Danaine Butterfly, *Idea leuconoe*. *Agric. Biol. Chem.* **1991**, *55*, 1787–1792. [CrossRef]
59. Hartmann, T.; Theuring, C.; Witte, L.; Schulz, S.; Pasteels, J.M. Biochemical processing of plant acquired pyrrolizidine alkaloids by the neotropical leaf-beetle *Platyphora boucardi*. *Insect Biochem. Mol. Biol.* **2003**, *33*, 515–523. [CrossRef]
60. Roitman, J.N. The pyrrolizidine alkaloids of *Amsinckia Menziesii*. *Aust. J. Chem.* **1983**, *36*. [CrossRef]
61. Stermitz, F.R.; Pass, M.A.; Kelley, R.B.; Liddell, J.R. Pyrrolizidine Alkaloids from Cryptantha Species. *Phytochemistry* **1993**, *33*, 383–387. [CrossRef]
62. Edgar, J.A.; Roeder, E.; Molyneux, R.J. Honey from Plants Containing Pyrrolizidine Alkaloids: A Potential Threat to Health. *J. Agric. Food Chem.* **2002**, *50*, 2719–2730. [CrossRef]
63. Gardner, D.R.; Thorne, M.S.; Molyneux, R.J.; Pfister, J.A.; Seawright, A.A. Pyrrolizidine alkaloids in *Senecio madagascariensis* from Australia and Hawaii and assessment of possible livestock poisoning. *Biochem. Syst. Ecol.* **2006**, *34*, 736–744. [CrossRef]
64. Atlas of Living Australia Website. Heliotropium amplexicaule Vahl. Available online: https://bie.ala.org.au/species/http://id.biodiversity.org.au/node/apni/2917623 (accessed on 1 August 2019).
65. Dubecke, A.; Beckh, G.; Lullmann, C. Pyrrolizidine alkaloids in honey and bee pollen. *Food Addit. Contam. Part. A* **2011**, *28*, 348–358. [CrossRef] [PubMed]
66. Fu, P.P.; Xia, Q.; Lin, G.; Chou, M.W. Pyrrolizidine alkaloids—genotoxicity, metabolism enzymes, metabolic activation, and mechanisms. *Drug Metab. Rev.* **2004**, *36*, 1–55. [CrossRef] [PubMed]
67. European Food Safety Authority. Dietary exposure assessment to pyrrolizidine alkaloids in the European population European Food Safety Authority (EFSA). *EFSA J.* **2016**, *14*, 4572. [CrossRef]
68. Crews, C.; Beuerle, T. The Analysis of Pyrrolizidine Alkaloids in Honey. In *Analysis of Food Toxins and Toxicants*; Wong, Y.-C., Lewis, R.J., Eds.; John Wiley & Sons Ltd.: Hoboken, NJ, USA, 2017; pp. 237–266.
69. Fu, P.P. Pyrrolizidine Alkaloids: Metabolic Activation Pathways Leading to Liver Tumor Initiation. *Chem. Res. Toxicol.* **2017**, *30*, 81–93. [CrossRef] [PubMed]

toxins

MDPI

Article

Development and Validation of a UHPLC-ESI-MS/MS Method for Quantification of Oleandrin and Other Cardiac Glycosides and Evaluation of Their Levels in Herbs and Spices from the Belgian Market

Svetlana V. Malysheva [1,*], Patrick P. J. Mulder [2] and Julien Masquelier [1]

1 Unit Toxins, Organic Contaminants and Additives, Sciensano, 1050 Brussels, Belgium; julien.masquelier@sciensano.be
2 Wageningen Food Safety Research, Wageningen University and Research, 6708 WB Wageningen, The Netherlands; patrick.mulder@wur.nl
* Correspondence: svetlana.malysheva@sciensano.be

Received: 6 March 2020; Accepted: 8 April 2020; Published: 9 April 2020

Abstract: Cardiac glycosides (CGs) are naturally occurring plant secondary metabolites that can be toxic to humans and animals. The aim of this work was to develop a targeted analytical method utilizing liquid chromatography—tandem mass spectrometry (LC-MS/MS) for quantification of these plant toxins in a herbal-based food and human urine. The method included oleandrin, digoxin, digitoxin, convallatoxin, and ouabain. Samples of culinary herbs were extracted with acetonitrile and cleaned using Oasis® MAX solid-phase extraction (SPE), while samples of urine were diluted with acidified water and purified on Oasis® HLB SPE cartridges. Limits of quantification were in the range of 1.5–15 ng/g for herbs and 0.025–1 ng/mL for urine. The mean recovery of the method complied with the acceptable range of 70–120% for most CGs, and relative standard deviations were at maximum 14% and 19% for repeatability and reproducibility, respectively. Method linearity was good with calculated R^2 values above 0.997. The expanded measurement uncertainty was estimated to be in the range of 7–37%. The LC-MS/MS method was used to examine 65 samples of culinary herbs and herb and spice mixtures collected in Belgium, from supermarkets and local stores. The samples were found to be free from the analyzed CGs.

Keywords: oleandrin; LC-MS/MS; plant toxins; validation; herbs; urine

Key Contribution: This work describes the development of a reliable analytical method for quantification of poisonous cardiac glycosides in food, namely culinary herbs and spices, and in human urine. Such methods are currently very scarce but paramount for investigation of accidental contamination with toxic plants through the alimentary chain or intentional poisoning with raw plant parts.

1. Introduction

Cardiac glycosides (CGs) are secondary metabolites produced by plants belonging to, among others, the genera *Nerium* (oleander), *Convallaria* (lily-of-the-valley), and *Digitalis* (foxglove). The core structure of most CGs consists of lactone and steroid rings and a sugar moiety (Figure 1). CGs occur in all parts of plants and can be poisonous to livestock and humans. Their primary mechanism of action is inhibition of the membrane sodium-potassium pump that influences the intracellular sodium, calcium, and potassium concentrations and, as a consequence, causes disruptions in the cardiovascular system. However, other symptoms of toxicity may also include gastrointestinal, ocular, and neurologic disorders. In a specific dose range, however, CGs such as digoxin (DIGO) and digitoxin (DIGI) (Figure 1), have a long history of use as medications in treating various heart conditions [1–6].

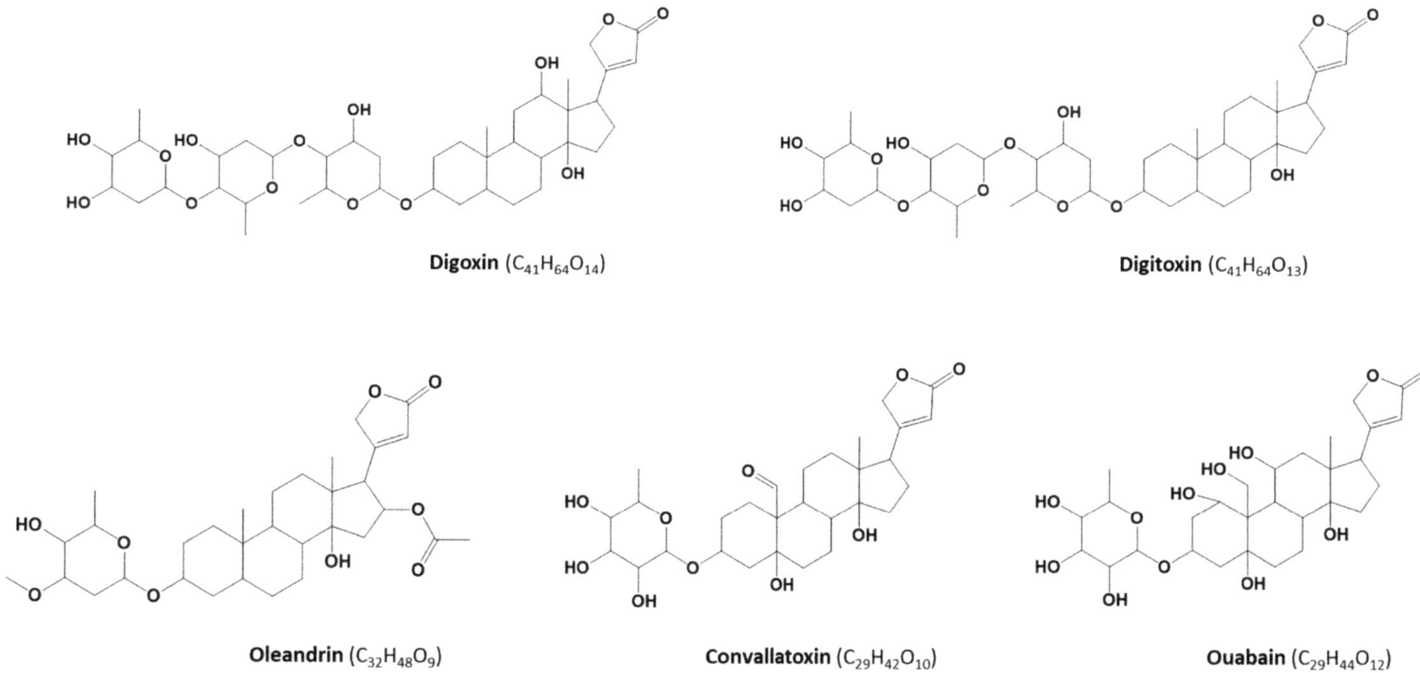

Figure 1. Chemical structures of the target cardiac glycosides.

In the literature, numerous cases of human poisoning with plants containing CGs through self-medication, accidental ingestion, suicide attempts, or criminal administration have been documented [7–11]. These reports also included a remarkable case of intoxication with CGs through food [8]. Superficial resemblance of the leaves of *Nerium oleander*, a plant producing toxic CG oleandrin (OLE), to the leaves of olive and bay trees (*Laurus nobilis*) might contribute to misidentification of the plant material and to accidental poisoning. Renal excretion is the main elimination route for some CGs (e.g., DIGO), while the hepatic route is more common to other CGs, combined with a partial renal elimination [12]. CGs excreted in urine are mainly unchanged [13–15] or partially metabolized [16].

A number of analytical, mostly single-analyte, methods for quantification of CGs in biological matrices have been described [11,17–24]. As a detection technique, these methods utilized mass spectrometry (MS) coupled to liquid chromatography (LC), which, thanks to its good selectivity and sensitivity, has nowadays become the method of choice for many applications including toxin analysis. Other techniques, such as immunoassay, (high-performance) thin-layer chromatography, and high-performance LC coupled to a UV or fluorescence detector [25–29], have also been applied to the detection and quantification of CGs.

Analytical methods for determination of these plant toxins in other than clinical samples are currently scarce. However, in case of a poisoning incident originating from the alimentary chain, the availability of a reliable method for food products is essential to confirm or rule out the ingestion of CG-containing plant material. Owing to the frequent use of herbal products as composite blends, and the complexity of the CG compound class, more than one CG might be associated with the poisoning, which points towards the significance of setting up multi-analyte methods. Therefore, the objective of this work was the development and validation of an ultra-high-performance (UHPLC)-MS/MS method for quantification of five plant toxins, namely OLE, DIGO, DIGI, convallatoxin (CON), and ouabain (OUB) (Figure 1), in edible herbs and spices and, complementary, in human urine. The choice of the target glycosides was dictated by their toxicity, known intoxication cases, occurrence of CG-producing plants, and availability of commercial reference standards.

2. Results

2.1. Optimization of LC-MS/MS Conditions

In this study, basic (pH 9, 10 mM ammonium bicarbonate (NH_4HCO_3) with ammonia (NH_3)) and acidic (pH 3, 10 mM ammonium formate ($HCOONH_4$) with formic acid (HCOOH)) aqueous mobile phases in combination with acetonitrile (ACN) as an organic phase were used to optimize the MS ionization of target CGs. As the initial step of optimization, the flow injection analysis was performed in electrospray ionization positive (ESI(+)) and negative (ESI(-)) modes. In the acidic mobile phase, the presence of intense $[M + Na]^+$ and $[M + K]^+$ adducts, which did not further fragment, was observed for all CGs; therefore, further optimizations were carried out with the basic mobile phase. Figure 2a,b demonstrates a full ESI(+)-MS spectrum of CON in the mobile phase at pH 3, in which an abundant presence of $[M + Na]^+$ and $[M + K]^+$ adducts can be clearly observed, and a full ESI(+)-MS spectrum in the basic mobile phase, in which the abundant presence of the molecular ion is apparent. As opposed to $[M + Na]^+$ and $[M + K]^+$ ions, fragmentation of the molecular ion provided several abundant product ions (Figure 2c), usable for defining selected reaction monitoring (SRM) transitions. While in the ESI(+) mode $[M + H]^+$ or $[M + NH_4]^+$ ions were abundant in the spectrum, the full MS scan in the ESI(-) mode revealed sufficient abundance of $[M-H]^-$ ions for all CGs. The final MS and MS/MS conditions (Table 1) were optimized in ESI(+), because of the higher intensity of MS signals in this mode.

For LC separation of the target analytes, the suitability of a UHPLC column with the C18 BEH stationary phase in combination with the basic mobile phase consisting of ACN and 10 mM NH_4HCO_3 at pH 9, was tested. Using a gradient elution (see Section 5.2), good separation of the CGs was achieved (Figure 3).

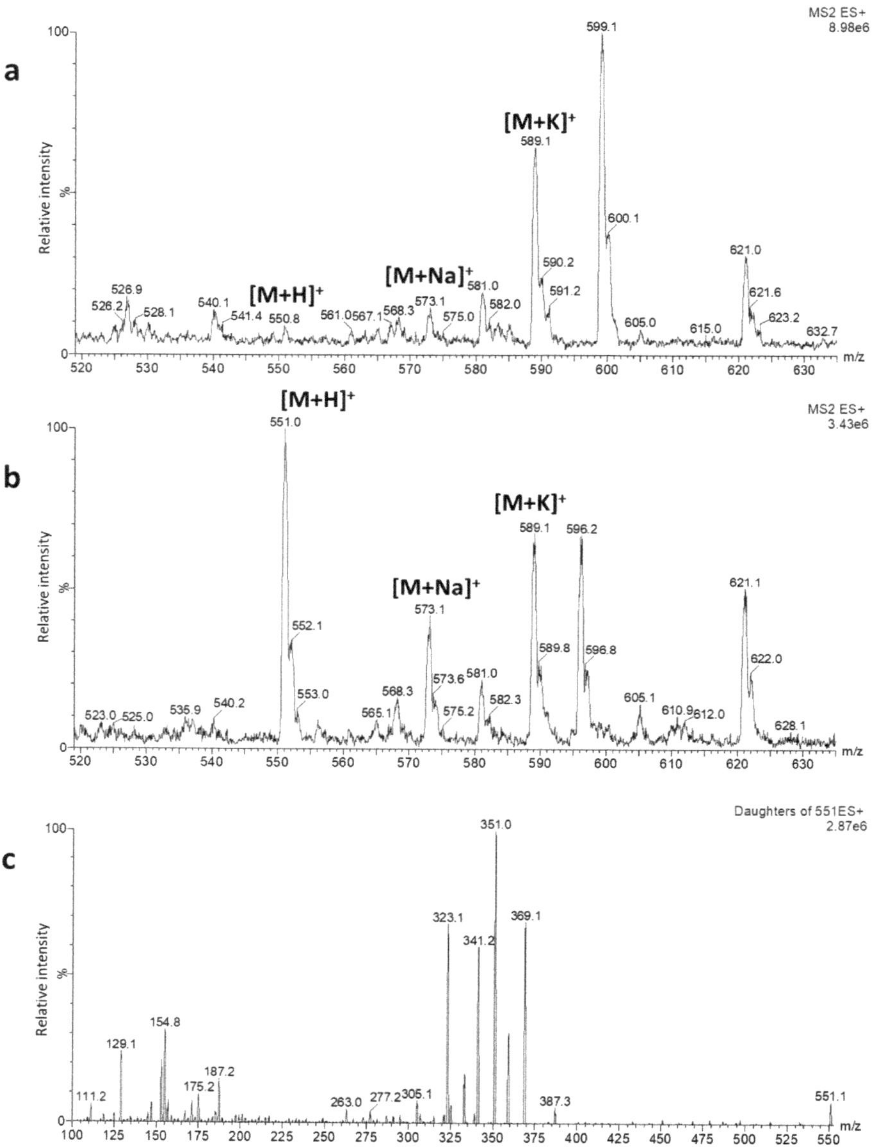

Figure 2. Full ESI(+)-MS spectra obtained through the flow injection analysis of a 1 µg/mL solution of convallatoxin (CON) in H_2O + 10 mM $HCOONH_4$ (pH 3):acetonitrile (ACN) (50:50, v/v) (**a**) and H_2O + 10 mM NH_4HCO_3 (pH 9):ACN (50:50, v/v) (**b**) and ESI(+)-MS/MS spectrum in H_2O + 10 mM NH_4HCO_3 (pH 9):ACN (50:50, v/v) (**c**). The vertical axes represent relative peak intensity (normalized to 100%), while the horizontal axes display measured *m/z* (mass-to-charge ratio) values. The MS setup is given in Section 5.3.

Table 1. Electrospray ionization positive (ESI)(+)-MS/MS parameters for detection of cardiac glycosides.

Analyte	Precursor Ion (*m*/*z*)	Cone Voltage (V)	Product Ions (*m*/*z*)	Collision Energy (eV)
Oleandrin	577.2 $[M + H]^+$	30	**373.2** [1] 433.1	15 10
Digoxin	781.2 $[M + H]^+$	25	**651.1** 391.1	10 15
Digitoxin	782.4 $[M + NH_4]^+$	30	**635.1** 375.1	10 20
Convallatoxin	551.1 $[M + H]^+$	20	**369.2** 351.2	10 20
Ouabain	585.1 $[M + H]^+$	30	**403.1** 385.1	15 20
Digoxin-d_3 [2]	784.2 $[M + H]^+$	25	**654.1** 394.2	10 15

[1] Highlighted in bold: Most abundant product ion, [2] internal standard.

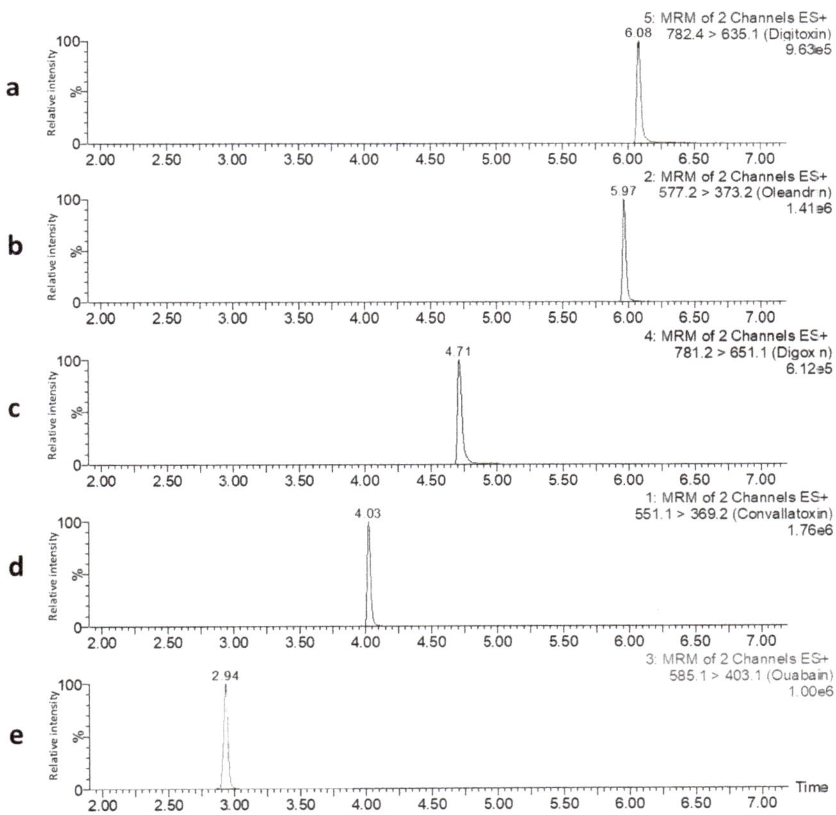

Figure 3. LC-MS/MS selected reaction monitoring (SRM) chromatograms of a single injection of a standard mixture of DIGI (**a**), DIGO (**c**), CON (**d**), and ouabain (OUB) (**e**) at a concentration of 2.5 ng/mL and OLE (**b**) at a concentration of 0.25 ng/mL, dissolved in H_2O:ACN (80:20, v/v). For each cardiac glycoside (CG) the most abundant SRM transition is displayed. The vertical axes represent relative peak intensity (normalized to 100%), while the horizontal axes display retention time (in min). The chromatographic conditions applied are given in Section 5.2.

2.2. Optimization of Sample Preparation

For the evaluation of extraction and clean-up recovery, analyte-free herbal and urine samples spiked with the target CGs before and after extraction and/or clean-up step were prepared.

A herbal mixture (herbes de Provence) was used as a test mixture for the extraction experiments. ACN, methanol (MeOH), and H_2O, as single solvents or as mixtures, were tested. It was found that good extraction recoveries (>70%) were obtained with ACN, MeOH, ACN:H_2O (50:50, v/v), and MeOH:H_2O (50:50, v/v), with slightly better results for OUB if MeOH was used in the extraction solvent. This method was aimed at achieving as low as possible limits of quantification (LOQs). It was apparent that to accomplish that a further clean-up and concentration step of the extract was necessary. The widely-used QuEChERS method [30], which combines extraction of a sample with ACN, salting-out, and subsequent dispersive solid phase extraction (SPE) clean-up, was tried, however, it demonstrated low extraction recoveries and poor clean-up efficiency. ENVI-Carb™ SPE with a graphite sorbent, which is very suitable for elimination of pigments that are abundantly present in herbs, resulted in no recovery of the target CGs. Other SPE cartridges, such as Discovery® DSC-18, and Oasis® HLB, provided acceptable recoveries but matrix effects were pronounced, affecting the sensitivity of the method. Oasis® MAX SPE was found to be the most suitable for clean-up of the herb samples, as it showed reduced matrix effects, good recoveries, and improved estimated LOQs. Among the tested extraction solvents, ACN was best compatible with the required setup of the Oasis® MAX protocol. Other solvents in combination with the herb matrix caused blockage of SPE cartridges or of the filter prior to the SPE. It should be mentioned that none of the tested protocols provided a sufficiently low LOQ and reproducible results for OUB, therefore, this compound was not included in the final method for herbal samples.

The sample preparation for urine was more straightforward and consisted of a sample dilution with H_2O containing 2% HCOOH and clean-up with Oasis® HLB SPE. The subsequent extract concentration was necessary to achieve a higher method sensitivity. In urine, this protocol was able to provide reproducible results for OUB and thus all selected plant toxins were included in the final method for this matrix.

The detailed protocols for extraction and clean-up of herb and urine samples are given in Section 5.4.

2.3. Method Validation

The method validation data for herbs and urine are summarized in Tables 2 and 3, respectively. The method LOQs were calculated based on a signal-to-noise ratio (S/N) approach and are reported in order to simplify the comparison with other methods for CGs described in the literature. The LOQs were in the range from 1.5 to 15 ng/g for herbs and from 0.025 to 1 ng/mL for urine, with OLE showing the highest sensitivity among the target CGs (Table S1). The analysis of blank herb and urine samples demonstrated that no peak with a S/N of at least 3 was detected at the expected retention time of the CGs, pointing out good specificity of the method. The matrix effect experiments revealed that the calculated *t*-value for the target CGs in herbs and urine were much greater than the tabulated *t*-value at the 95% confidence level indicating a significant difference between the slopes of calibration curves in the solvent and matrix, i.e., the presence of matrix effects. All CGs in both matrices suffered from a signal suppression, with the strongest suppression for OUB and the smallest effect for OLE and DIGI. The calibration curves were prepared in matrix extracts (spiked post clean-up) by plotting the concentration of the analyte in the calibration standards against the ratio of peak area of the analyte to the internal standard, digoxin-d_3 (DIGO-D), for all target CGs. Calibration curves for herb and urine samples were linear over the validated concentration range with coefficients of determination (R^2) >0.997. The lowest calibration level (LCL) is used as a reporting limit for quantification of CGs in herbs and urine. The mean (apparent) recovery data obtained for three concentration levels in herbs were in the range from 83% to 115% for OLE, DIGO, and DIGI, and 55% for CON. The mean recoveries for all CGs in urine ranged from 80% to 96%. The method precision was expressed as a relative standard deviation (RSD) of replicate measurements. For herbs, the repeatability (RSD_r) of the method ranged from 6% to 14% and the within-laboratory reproducibility (RSD_{wR}) was from 7% to 17%, while these parameters for urine ranged from 1% to 7% and from 5% to 19%, respectively. The expanded measurement uncertainty (MU) was not higher than 28% and 37% at the lowest concentration levels validated for herbs and urine, respectively. The uncertainty at higher concentration levels did not exceed 31% and 16% for herbs and urine, respectively.

2.4. Method Application for Analysis of Culinary Herbs

The validated LC-MS/MS method was subsequently used to investigate the contamination of culinary herbs and spices that are available on the Belgian food market. In total, 65 samples were acquired in supermarkets and organic food shops and comprised the culinary herbs and herb/spice mixtures containing bay leaves (*Laurus nobilis*). For the majority of samples, the country of production was not specified. About 20% of samples originated from organic farming. The detailed information on ingredients of the samples is given in Table S2.

Quality control samples, namely a standard mixture of CGs in a neat solvent and herb mixture fortified with CGs at the concentrations corresponding to the middle validated level, were included in each sample sequence. Identification of CGs in samples was completed following the Commission Decision 2002/657/EC [31]. This implied the presence of a peak of the target analyte with a S/N ratio of at least 3 for each ion transition, compliance of relative retention times, and conformity of deviations of relative ion intensities with regards to the matrix-matched calibration standards. The most abundant product ion was used for quantification, while the second product ion was used for confirmation of the analytes. The analysis demonstrated that none of the collected samples contained CGs above the LCL.

Table 2. Validation data for cardiac glycosides in culinary herbs.

Analyte	Reporting Limit (ng/g)	Linear Range (ng/g)	R^2	Concentration Level (ng/g)	Recovery ± SD[1] (%)	Repeatability (RSD_r) (%)	Reproducibility (RSD_{wR}) (%)	Measurement Uncertainty (%)
Oleandrin	2.5	2.5–200	0.9993	5	82 ± 7	7	8	17
				25	86 ± 6	6	7	14
				100	80 ± 6	6	7	15
					Mean: 83 ± 6	Mean: 6	Mean: 8	
Digoxin	10	10–400	0.9995	10	121 ± 11	11	11	21
				50	114 ± 7	7	7	13
				200	111 ± 10	6	10	19
					Mean: 115 ± 10	Mean: 9	Mean: 9	
Digitoxin	20	20–400	0.9995	20	86 ± 10	14	14	27
				100	96 ± 11	9	12	24
				400	95 ± 13	7	15	31
					Mean: 92 ± 12	Mean: 10	Mean: 14	
Convallatoxin	10	10–400	0.9994	10	52 ± 7	7	14	28
				50	55 ± 9	12	17	34
				200	57 ± 7	10	12	25
					Mean: 55 ± 7	Mean: 9	Mean: 14	

[1] SD: Standard deviation.

Table 3. Validation data for cardiac glycosides in human urine.

Analyte	Reporting Limit (ng/mL)	Linear Range (ng/mL)	R^2	Concentration Level (ng/mL)	Recovery ± SD[1] (%)	Repeatability (RSD_r) (%)	Reproducibility (RSD_{wR}) (%)	Measurement Uncertainty (%)
Oleandrin	0.1	0.1–20	0.9987	0.5	80 ± 13	3	18	36
				2.5	84 ± 9	3	13	26
				10	87 ± 5	2	7	13
					Mean: 84 ± 10	Mean: 3	Mean: 13	
Digoxin	1	1–200	0.9993	5	93 ± 10	4	12	24
				25	92 ± 6	2	7	14
				100	92 ± 4	3	5	9
					Mean: 92 ± 7	Mean: 3	Mean: 8	
Digitoxin	1	1–200	0.9975	5	86 ± 14	7	19	37
				25	95 ± 10	6	12	24
				100	88 ± 5	3	5	11
					Mean: 89 ± 11	Mean: 5	Mean: 12	
Convallatoxin	1	1–200	0.9993	5	91 ± 15	4	19	36
				25	94 ± 7	2	8	26
				100	96 ± 3	1	3	16
					Mean: 94 ± 9	Mean: 3	Mean: 10	
Ouabain	1	1–200	0.9994	5	87 ± 14	4	18	37
				25	89 ± 11	6	13	17
				100	93 ± 7	4	8	7
					Mean: 90 ± 11	Mean: 5	Mean: 13	

[1] SD: Standard deviation.

3. Discussion

Phytotherapeutic and nutritional use of plants and herbal-based products has (re)gained its popularity in the last years. Due to improper usage of plants or unawareness of plant toxicity, several intoxication cases with CG-containing plants have been reported recently [7,10,11]. Ingestion of toxic plant (parts) may also be possible through the food chain. Plant misidentification and inadequate control of harvesting or processing may lead to an unintentional mix of toxic plant material with the raw plant material used for production. Since more analytical methods became available for screening of organic molecules at low levels, several contaminants have surfaced as an issue of relevance in food safety. That was the case for other plant toxins, such as tropane alkaloids and pyrrolizidine alkaloids, which were found in herbal teas, herbs and spices, cereal-based food, and herbal food supplements [32–35], sometimes at levels that can represent risk for human health [36,37]. The current study aimed at developing a reliable analytical method for detection of CGs in plant-based food products. As a complementary tool for control of poisoning incidents, the method was also validated for the urine matrix. As opposed to some other human biological fluids that are used to study exposure to contaminants, urine is easily accessible from individuals of all ages, can be obtained in larger volumes, and its collection method is noninvasive.

As a detection technique, LC-MS/MS was chosen for this work, as it allows a high-throughput simultaneous detection of structurally diverse molecules, including compounds of natural origin, at trace levels and with high selectivity and specificity. In the proposed method, the SRM mode was used to obtain increased sensitivity and specificity, the parameters that are of great importance in the analysis of such complex matrices as herbs and urine. The selected ionization mode was ESI, the commonly used interface in the LC-MS analysis of natural toxins. Though less frequent, atmospheric pressure chemical ionization (APCI) is also used for the ionization of small molecules. Sugergat et al. [38] compared these two ionization modes for the analysis of DIGO in human serum and observed a lower intensity of the protonated molecule and a higher degree of fragmentation, resulting in lower sensitivity of the APCI mode compared to ESI.

For some molecules, formation of alkali metal adducts can be observed in ESI-MS. This can possibly be attributed to leaching from glass recipients or the presence of impurities in the mobile phase [39]. Such adducts, as compared to the molecular ions, can be unstable and might not produce fragment ions, jeopardizing the reliability of a quantitative LC-MS/MS method. The particular issue of metal adduct formation has also been reported for CGs [18]. In this study, formation of metal adducts, not prone to fragmentation, was observed when a mobile phase with $HCOONH_4$ and HCOOH (pH 3) was used for the LC. In order to generate a sensitive single precursor ion for the analysis in the SRM mode, Bylda et al. investigated different mobile phase additives and finally selected the $[M + Li]^+$ adduct for the quantification of CGs [18]. In the absence of intense molecular ions, some applications used $[M + Na]^+$ or $[M + K]^+$ adducts for quantification of CGs in a single or selected ion monitoring (SIM) mode [38,40]. However, it has been noticed that SIM produced much higher detection limits with biological samples compared to SRM, while the two modes were similar in sensitivity if a standard mixture of CGs containing no matrix was injected [21]. In the current method, under the applied LC-MS conditions (see Section 5.2) with a NH_4HCO_3-containing mobile phase (pH 9), formation of a protonated molecular ion $[M + H]^+$ for DIGO, OLE, OUB, and CON, and of an $[M + NH_4]^+$ adduct for DIGI was achieved. Under mild fragmentation conditions, these precursor ions underwent collision-induced dissociation in the quadrupole yielding usable intense product ions. Of these fragments, the two most abundant ions were selected for the identification and quantification of these plant toxins, thereby fulfilling the requirements of the Commission Decision 2002/657/EC, which recommends the use of four identification points for confident identification of compounds in the LC-MS/MS analysis [31]. Some of the observed abundant product ions corresponded to sequential losses of the sugar moieties and elimination of hydroxyl groups from the steroid aglycones [41].

To optimize the sensitivity and selectivity of the LC-MS/MS method, SPE was used in the sample preparation of CGs in herbs and urine, in order to reduce complexity of the matrix and to allow enrichment of the analytes. This approach has previously been used for purification of some CGs in whole blood and plasma [17,24,42]. Compared to Oasis® HLB SPE devices, used for the clean-up of urine samples in the current method, a significant reduction of matrix effects for herb samples was noted with Oasis® MAX cartridges.

Currently, no maximum levels for CGs in food are set and analytical methods should preferably achieve as low as possible LOQs. The results of validation indicated that the developed method is able to detect and quantify the target CGs at low levels. In comparison to other existing methods for CGs in biological matrices and herbal-based products (Table S1), the current method is able to reach lower or similar LOQs. During the validation, matrix effects were observed for herbs and urine. The matrix effect is in many occasions unavoidable in MS analysis of complex matrices. It could be caused by co-eluting compounds that interfere with the ionization process of the target analytes leading to signal suppression or enhancement. The CGs were differently affected by the matrix interferences. OUB, being the most polar CG and eluting early in the chromatographic run, suffered from a greater matrix suppression than OLE and DIGI, which eluted at the end of the run. This can be associated with the fact that hydrophilic compounds of the biological or plant matrix also eluted early in the chromatographic run giving rise to a more pronounced matrix effect in this region of the chromatogram. The addition of stable isotope labeled analogues of the analyzed molecules is suggested for use as internal standards to counteract the matrix effects. The use of commercially available or in-house synthesized labeled internal standards have already been reported for the LC-MS analysis of a number of CGs [18,19]. Other structurally related molecules, such as methyldigoxin, digitoxigenin, and gitoxigenin, were included as internal standards in some analytical methods [21,40,43]. In the current method DIGO-D, the deuterated analogue of DIGO, was applied as an internal standard for all target CGs.

The method linearity was tested and found to be good exhibiting R^2 >0.997, in most cases >0.999. The mean recovery of the developed method was in agreement with the acceptable limit of 70–120%, except for CON in herbs. A good precision of the method was demonstrated with repeatability and within-laboratory reproducibility below 20% for both matrices. The accuracy and the precision of the developed method were in the same range as for other methods, reported in the literature, for quantification of CGs in biological matrices (Table S1). The obtained MU was below 50%, complying with the SANTE/12682/2019 guidance document [44].

After completing the validation, the developed LC-MS/MS method was applied, to assess the CG contamination of herbs and herbal/spice mixtures used for culinary purposes, available on the Belgian food market. None of the target CGs were detected in these samples indicating that there is currently no safety risk for the population with regards to contamination of culinary herbs and blends with CGs.

CGs are a large and diverse group of naturally occurring toxic compounds, and, upon necessity, the LC-MS technique allows extension of the method to other plant toxins from this class. The presented method can, on the one hand, be used in food control initiatives to ensure food safety and, on the other hand, in population-wide survey studies that combine monitoring of food contamination and analysis of human biological fluids to unravel the level of exposure to plant toxins.

4. Conclusions

This study describes the first detailed validated method for quantification of OLE and other CGs in culinary herbs and human urine. The method displays good specificity, linearity, accuracy, and expanded measurement uncertainty, thus enabling the accurate quantification of OLE, DIGO, DIGI, and CON in culinary herbs and OLE, DIGO, DIGI, CON, and OUB in human urine. This new method was applied to the analysis of more than 60 samples of culinary herbs and herb/spice mixtures containing bay leaves present on the Belgian food market, showing that these products are safe for the consumer. The UHPLC-ESI-MS/MS method described here could, therefore, become a useful tool to determine these plant toxins in culinary herbs and also in urine.

5. Materials and Methods

5.1. Standards, Reagents, and Consumables

Analytical standards of OLE, DIGO, DIGI, CON, and OUB octahydrate were purchased from Sigma-Aldrich (Buchs, Switzerland). Individual stock solutions were prepared by dissolving the crystalline standards in MeOH at a concentration of 1 mg/mL. A methanolic solution of DIGO-D (1 mg/mL) was obtained from Cayman Chemicals (Ann Arbor, MI, USA). Intermediate solutions were prepared by diluting the stock solutions in MeOH. The stock and the intermediate solutions were stored at −20 °C.

The MeOH absolute ULC-MS, ACN ULC-MS, and HCOOH 99% ULC-MS were purchased from Biosolve (Valkenswaard, the Netherlands). The ammonia solution 28–30% was obtained from Merck (Darmstadt, Germany), while NH_4HCO_3 LC-MS and $HCOONH_4$ LC-MS were supplied by Sigma-Aldrich (Steinheim, Germany). H_2O was purified by a Milli-Q purification system (Millipore Corp., Bedford, MA, USA).

Oasis® MAX (3 cc, 60 mg) LP and Oasis® HLB (3 cc, 60 mg) extraction cartridges were provided by Waters (Wexford, Ireland). Discovery® DSC-18 (6 mL, 500 mg) and Supelclean™ ENVI-Carb™ (6 mL, 500 mg) SPE cartridges were obtained from Sigma-Aldrich. VWR (Randor, PA, USA) was the supplier of 15 and 50 mL centrifuge PP tubes and centrifugal filters (modified nylon, 0.2 µm, 500 µL).

5.2. UHPLC-MS/MS Conditions

The UHPLC-MS/MS system consisted of an ACQUITY UPLC H-class system coupled to a Xevo TQ-S triple quadrupole mass spectrometer (Waters, Milford, MA, USA).

The mass spectrometer was operated in the ESI(+) mode. The MS parameters were set as follows: Source and desolvation temperatures: 150 and 350 °C, respectively; capillary voltage: 1.50 kV; cone and desolvation gas flows: 150 and 1000 L/h, respectively; collision gas flow: 0.15 mL/min; source offset: 30 V. The SRM acquisition mode was used.

Chromatographic separation was achieved on an AQUITY UPLC™ BEH C18 column (2.1 × 100 mm; 1.7 µm) with an ACQUITY UPLC™ BEH C18 VanGuard precolumn (2.1 × 5 mm; 1.7 µm) (both from Waters). The column temperature was maintained at 40 °C. The mobile phase was composed of phase A (H_2O containing 10 mM NH_4HCO_3 pH 9) and phase B (ACN). The flow rate used was 0.45 mL/min and the applied gradient elution program was as follows: 0–1 min: 95% A, 1–6 min: 40% A, 6–7 min: 10% A, 7–7.1 min: 95% A, 7.1–10 min: 95% A. The injection volume was 10 µL.

5.3. Samples

The samples of culinary herbs were purchased from supermarkets and local (organic food) stores in Belgium and included 14 single herbs and 51 herb/spice mixtures. More details on the samples are provided in Table S2. Prior to the analysis, the herb samples were finely ground and homogenized. The samples of human urine were provided by volunteers.

5.4. Sample Preparation

Two grams of the herb sample was weighed in a 50 mL PP tube. After addition of 25 mL of ACN, the sample was vigorously shaken on an overhead shaker for 30 min and centrifuged for 10 min at 3180× *g*. Ten mL of supernatant was transferred in a 15 mL PP tube and evaporated at 45 °C under a stream of nitrogen until a volume of approximately 1 mL. Subsequently, H_2O containing 5% NH_3 was added to a total volume of 10 mL. After thorough vortexing and centrifugation (10 min at 3180× *g*) the extract was subjected to further clean-up. Oasis® MAX SPE cartridges were conditioned with 3 mL MeOH and 3 mL H_2O. Three mL of supernatant was loaded onto the cartridge and washed with 3 mL H_2O containing 5% NH_3. After a brief drying step, the target analytes were eluted with 3 mL MeOH and collected in 15 mL PP tubes. The eluate was evaporated until dryness at 45 °C under a stream of nitrogen. The residue was reconstituted in 250 μL H_2O:ACN (80:20, v/v) and filtered through filter units for 5 min at 14,000× *g*.

Five mL of urine was transferred to a 15 mL PP tube, to which 5 mL H_2O containing 2% HCOOH was added. After vortexing, the sample was centrifuged for 10 min at 3180× *g*. For the clean-up, Oasis® HLB cartridges were conditioned with 3 mL MeOH and 3 mL H_2O. Six mL of supernatant was loaded onto the cartridge and washed with 3 mL H_2O. After a brief drying step, the target analytes were eluted with 3 mL MeOH and collected in 15 mL PP tubes. The eluate was evaporated until dryness at 45 °C under a stream of nitrogen. The residue was reconstituted in 500 μL H_2O:ACN (80:20, v/v) and filtered through filter units for 5 min at 14,000× *g*.

5.5. Validation

The validation study was performed using spiked analyte-free representative sample materials. The following method parameters were evaluated: LOQ, specificity, linearity, recovery, repeatability (RSD_r), reproducibility (RSD_{wR}), matrix effects, and expanded measurement uncertainty (MU).

LOQ was defined as the minimum analyte concentration in the spiked samples that produced an SRM transition with a minimum S/N of 10. Specificity of the method was assessed through the analysis of blank matrix samples. The absence of a peak with a S/N ≥ 3 at the expected retention time of the target CG indicated good method specificity. Linearity of the method was evaluated by fortifying blank matrix samples with the target analytes at varying concentrations (minimum five levels). A logarithmic transformation of the axes and a linear regression model were applied. (Apparent) recovery was assessed by a spiking of blank matrix with the target analytes at three concentration levels in triplicate. The measured concentrations were determined using a matrix-matched calibration curve and the recovery was calculated as follows (Equation 1):

$$\text{Recovery (\%)} = \text{Measured concentration/Theoretical concentration} \times 100. \quad (1)$$

For determination of the repeatability (RSD_r), spiking experiments were performed at three concentration levels in triplicate on the same day, while for within-laboratory reproducibility (RSD_{wR}) evaluation, the same experiments were carried out on three days. Matrix effects were assessed by comparing the slopes of calibration curves prepared in the matrix extract and neat solvent. The *t*-test was used for statistical evaluation of the matrix effect data. The expanded measurement uncertainty (MU) (corresponding to a 95% confidence level and a coverage factor of 2) was estimated according to [44].

Supplementary Materials: The following are available online at http://www.mdpi.com/2072-6651/12/4/243/s1, Table S1: Comparison of performance characteristics of different LC-MS/MS methods for analysis of cardiac glycosides; Table S2: Overview of culinary herbs and spices collected in Belgian food stores.

Author Contributions: Conceptualization, S.V.M. and J.M.; methodology, S.V.M., P.P.J.M., and J.M.; validation, S.V.M. and J.M.; formal analysis, S.V.M. and J.M.; investigation, S.V.M., P.P.J.M., and J.M.; data curation, S.V.M.; writing—original draft preparation, S.V.M., P.P.J.M., and J.M.; writing—review and editing, S.V.M., P.P.J.M., and J.M.; visualization, S.V.M., P.P.J.M., and J.M.; supervision, J.M.; project administration, S.V.M. All authors have read and agreed to the published version of the manuscript.

Funding: This research was funded by the Belgian National Reference Laboratory for Mycotoxins, Plant Toxins, and Marine Biotoxins (contract number FAVV-NRL-GGO-TOX-2017, lot 2).

Acknowledgments: Benoît Guillaume and Céline Segers are acknowledged for their technical support, and Séverine Goscinny is thanked for providing a part of the herb samples.

Conflicts of Interest: The authors declare no conflict of interest. The funders had no role in the design of the study; in the collection, analyses, or interpretation of data; in the writing of the manuscript, or in the decision to publish the results.

References

1. Fozzard, H.A.; Sheets, M.F. Cellular mechanism of action of cardiac glycosides. *J. Am. Coll. Cardiol.* **1985**, *5*, 10A–15A. [CrossRef]
2. Hollman, A. Plants and cardiac glycosides. *Br. Heart J.* **1985**, *54*, 258–261. [CrossRef] [PubMed]
3. Melero, C.P.; Medarde, M.; San Feliciano, A. A short review on cardiotonic steroids and their aminoguanidine analogues. *Molecules* **2000**, *5*, 51–81. [CrossRef]
4. Kanji, S.; MacLean, R.D. Cardiac glycoside toxicity More than 200 years and counting. *Crit. Care Clin.* **2012**, *28*, 527–535. [CrossRef]
5. Morsy, N. Cardiac glycosides in medicinal plants. In *Aromatic and Medicinal Plants (Back to Nature)*; El-Shemy, H.A., Ed.; InTechOpen: London, UK, 2017. [CrossRef]
6. Botelho, A.F.M.; Pierezan, F.; Soto-Blanco, B.; Melo, M.M. A review of cardiac glycosides: Structure, toxicokinetics, clinical signs, diagnosis and antineoplastic potential. *Toxicon* **2019**, *158*, 63–68. [CrossRef]
7. Wasfi, I.A.; Zorod, O.; Al katheeri, N.A.; Al Awadhi, A.M. A fatal case of oleandrin poisoning. *Forensic. Sci. Int.* **2008**, *179*, e31–e36. [CrossRef]
8. Gechtman, C.; Guidugli, F.; Marocchi, A.; Masarin, A.; Zoppi, F. Unexpectedly dangerous escargot stew: Oleandrin poisoning through the alimentary chain. *J. Anal. Toxicol.* **2006**, *30*, 683–686. [CrossRef]
9. Hugues, T.; Anoult, M.; Beaub, N.; Yaici, K.; Mélandri, P.; Saoudi, N.; Gibelin, P. Intoxication volontaire au laurier rose; cas clinique et revue de la littérature. A non fatal *Nerium oleander* self-poisoning; case report and discussion. *Ann. Cardiol. Angeiol.* **2012**, *61*, 128–131. [CrossRef]
10. Papi, L.; Luciani, A.B.; Forni, D.; Guisiani, M. Unexpected double lethal oleander poisoning. *Am. J. Forensic. Med. Pathol.* **2012**, *33*, 93–97. [CrossRef]
11. Bavunoğlu, I.; Balta, M.; Türkmen, Z. Oleander poisoning as an example of self-medication attempt. *Balkan. Med. J.* **2016**, *33*, 559–562. [CrossRef]
12. Smith, T.W. Pharmacokinetics, bioavailability and serum levels of cardiac glycosides. *J. Am. Coil. Cordiol.* **1985**, *5*, 43A–50A. [CrossRef]
13. Doherty, J.E.; Kane, J.J. Clinical pharmacology of digitalis glycosides. *Annu. Rev. Med.* **1975**, *26*, 159–171. [CrossRef] [PubMed]
14. Lahrtz, H.g.; Reinold, H.M.; van Zwieten, P.A. Serum concentration and urinary excretion of ^{3}H-ouabain in patients suffering from liver or kidney diseases. *Pharmacol. Clinica.* **1969**, *1*, 114–118. [CrossRef]
15. Lisalo, E. Clinical pharmacokinetics of digoxin. *Clin. Pharm.* **1977**, *2*, 1–16. [CrossRef]
16. Hinderling, P.H.; Garrett, E.R.; Wester, R.C. Pharmacokinetics of b-methyldigoxin in healthy humans I: Intravenous studies. *J. Pharm. Sci.* **1977**, *66*, 242–253. [CrossRef]
17. Baecher, S.; Kroiss, M.; Fassnacht, M.; Vogeser, M. No endogenous ouabain is detectable in human plasma by ultra-sensitive UPLC-MS/MS. *Clin. Chim. Acta* **2014**, *431*, 87–92. [CrossRef]
18. Bylda, C.; Thiele, R.; Kobold, U.; Volmer, D.A. Simultaneous quantification of digoxin, digitoxin, and their metabolites in serum using high performance liquid chromatography-tandem mass spectrometry. *Drug Test. Anal.* **2015**, *7*, 937–946. [CrossRef]
19. Frommherz, L.; Köhler, H.; Brinkmann, B.; Lehr, M.; Beike, J. LC-MS assay for quantitative determination of cardio glycoside in human blood samples. *Int. J. Legal. Med.* **2008**, *122*, 109–114. [CrossRef]
20. Grabowski, T.; Świerczewska, A.; Borucka, B.; Sawicka, R.; Sasinowska-Motyl, M.; Gumułka, S.W.; Zahariev, I.; Mitova, A.; Zhilkova, K. Rapid chromatographic/mass spectrometric method for digoxin quantification in human plasma. *Pharm. Chem. J.* **2009**, *43*, 710–715. [CrossRef]
21. Guan, F.; Ishii, A.; Seno, H.; Watanabe-Suzuki, K.; Kumazawa, T.; Suzuki, O. Identification and quantification of cardiac glycosides in blood and urine samples by HPLC/MS/MS. *Anal. Chem.* **1999**, *71*, 4034–4043. [CrossRef]
22. Mitamura, K.; Horikawa, A.; Yamane, Y.; Ikeda, Y.; Fujii, Y.; Shimada, K. Determination of digoxin in human serum using stable isotope dilution liquid chromatography/electrospray ionization-tandem mass spectrometry. *Biol. Pharm. Bull.* **2007**, *30*, 1653–1656. [CrossRef] [PubMed]
23. Zhai, J.X.; Yan, H.; Shen, M.; Shen, B.H.; Liu, W. Determination of oleandrin in blood and liver samples by LC-MS/MS. *Fa Yi Xue Za Zhi* **2018**, *34*, 585–589. [CrossRef]

24. Gosetti, F.; Nebbia, C.; Ceci, L.; Carelli, G.; Marengo, E. UHPLC-MS/MS determination of oleandrin in blood and tissues of dairy cattle poisoned by oleander (*Nerium oleander*). *Anal. Methods* **2019**, *11*, 5562–5567. [CrossRef]
25. Dasgupta, A.; Datta, P. Rapid detection of oleander poisoning using digoxin immunoassays: Comparison of five assays. *Ther. Drug Monit.* **2004**, *26*, 658–663. [CrossRef] [PubMed]
26. Galey, F.D.; Holstege, D.M.; Plumlee, K.H.; Tor, E.; Johnson, B.; Anderson, M.L.; Blanchard, P.C.; Brown, F. Diagnosis of oleander poisoning in livestock. *J. Vet. Diagn. Investig.* **1996**, *8*, 358–364. [CrossRef]
27. Hamada, K.; Iwamoto, A.; Miyazaki, S.; Yamanaka, N.; Guruge, K.S. Determination of bovine blood oleandrin by high-performance liquid chromatography and postcolumn derivatization. *J. Chromatogr. Sci.* **2002**, *40*, 515–518. [CrossRef]
28. Namera, A.; Yashiki, M.; Okada, K.; Iwasaki, Y.; Kojima, T. Rapid quantitative analysis of oleandrin in human blood by high-performance liquid chromatography. *Nihon. Hoizaku. Zasshi.* **1997**, *51*, 315–318.
29. Praveen, U.S.; Gowtham, M.D.; Yogaraje-Gowda, C.V.; Nayak, V.G.; Mohan, B.M. Detection of residues of cardenolides of *Nerium oleander* by high-performance thin-layer chromatography in autopsy samples. *Int. J. Med. Toxicol. Forensic. Med.* **2012**, *2*, 135–142.
30. Anastassiades, M.; Lehotay, S.J.; Stajnbaher, D.; Schenck, F.J. Fast and easy multiresidue method employing acetonitrile extraction/partiotioning and "dispersive solid-phase extraction" for the determination of pesticide residues in produce. *J. AOAC Int.* **2003**, *86*, 412–431. [CrossRef]
31. Commission Decision of 12 August 2002 implementing Council Directive 96/23/EC concerning the performance of analytical methods and the interpretation of the results (2002/657/EC). *OJ* **2002**, *L221*, 8–36. Available online: https://eur-lex.europa.eu/legal-content/EN/TXT/PDF/?uri=CELEX:32002D0657&from=EN (accessed on 16 January 2020).
32. Mulder, P.P.J.; De Nijs, M.; Castellari, M.; Hortos, M.; MacDonald, S.; Crews, C.; Hajslova, J.; Stranska, M. Occurrence of tropane alkaloids in food. *EFSA Support. Publ.* **2016**, EN-1140. [CrossRef]
33. Mulder, P.P.J.; López, P.; Castellari, M.; Bodi, D.; Ronczka, S.; Preiss-Weigert, A.; These, A. Occurrence of pyrrolizidine alkaloids in animal- and plant-derived food: Results of a survey across Europe. *Food Addit. Contam. Part A* **2018**, *35*, 118–133. [CrossRef] [PubMed]
34. Picron, J.-F.; Herman, M.; Van Hoeck, E.; Goscinny, S. Analytical strategies for the determination of pyrrolizidine alkaloids in plant based food and examination of the transfer rate during the infusion process. *Food Chem.* **2018**, *266*, 514–523. [CrossRef] [PubMed]
35. Kaltner, F.; Rychlik, M.; Gareis, M.; Gottschalk, C. Occurrence and risk assessment of pyrrolizidine alkaloids in spices and culinary herbs from various geographical origins. *Toxins* **2020**, *12*, 155. [CrossRef] [PubMed]
36. Knutsen, H.K.; Alexander, J.; Barregård, L.; Bignami, M.; Brüschweiler, B.; Ceccatelli, S.; Cottrill, B.; Dinovi, M.; Edler, L.; Grasl-Kraupp, B.; et al. Statement on the risks for human health related to the presence of pyrrolizidine alkaloids in honey, tea, herbal infusions and food supplements. *EFSA J.* **2017**, *15*, 4908–4942. [CrossRef]
37. Arcella, D.; Altieri, A.; Horváth, Z. Scientific report on human acute exposure assessment to tropane alkaloids. *EFSA J.* **2018**, *16*, 5160–5189. [CrossRef]
38. Sugergat, H.; Unger, K.K.; Emmert, J.; Wendt, J.; Mandel, F. Determination of digoxin in human serum by LC-MS with on-line sample preparation. *Appl. Notes Agil. Technol.* **2001**, *15*, 5988-364EN.
39. Nozaki, K.; Tarui, A.; Osaka, I. Elimination technique for alkali metal ion adducts from an electrospray ionization process using an on-line ion suppressor. *Anal. Sci.* **2010**, *26*, 715–718. [CrossRef]
40. Gozalpour, E.; Greupink, R.; Bilos, A.; Verweij, V.; van den Heuvel, J.J.M.W.; Masereeuw, R.; Russel, F.G.M.; Koenderink, J.B. Convallatoxin: A new P-glycoprotein substrate. *Eur. J. Pharmacol.* **2014**, *744*, 18–27. [CrossRef]
41. Ravi, B.G.; Guardian, M.G.E.; Dickman, R.; Wang, Z.Q. Profiling and structural analysis of cardenolides in two species of *Digitalis* using liquid chromatography coupled with high-resolution mass spectrometry. *J. Chromatogr. A* **2020**, in press. [CrossRef]
42. Carlier, J.; Guitton, J.; Romeuf, L.; Bévalot, F.; Boyer, B.; Fanton, L.; Gaillard, Y. Screening approach by ultra-high performance liquid chromatography-tandem mass spectrometry for the blood quantification of thirty-four toxic principles of plant origin. Application to forensic toxicology. *J. Chromatogr. B* **2015**, *975*, 65–76. [CrossRef] [PubMed]

43. Kanno, S.; Watanabe, K.; Yamagishi, I.; Hirano, S.; Minakata, K.; Gonmori, K.; Suzuki, O. Simultaneous analysis of cardiac glycosides in blood and urine by thermoresponsive LC-MS-MS. *Anal. Bioanal. Chem.* **2011**, *399*, 1141–1149. [CrossRef] [PubMed]
44. SANTE/12682/2019. *Analytical Quality Control and Method Validation Procedures for Pesticide Residues Analysis in Food and Feed*; European Commission: Brussels, Belgium, 2020; p. 49.

toxins

MDPI

Article

Revisiting the Neuroblastoma Cell-Based Assay (CBA-N2a) for the Improved Detection of Marine Toxins Active on Voltage Gated Sodium Channels (VGSCs)

Jérôme Viallon, Mireille Chinain and Hélène Taiana Darius *

Institut Louis Malardé (ILM), Laboratory of Marine Biotoxins-UMR 241-EIO, 98713 Papeete-Tahiti, French Polynesia; jviallon@ilm.pf (J.V.); mchinain@ilm.pf (M.C.)

* Correspondence: tdarius@ilm.pf

Received: 3 April 2020; Accepted: 24 April 2020; Published: 27 April 2020

Abstract: The neuroblastoma cell-based assay (CBA-N2a) is widely used for the detection of marine biotoxins in seafood products, yet a consensus protocol is still lacking. In this study, six key parameters of CBA-N2a were revisited: cell seeding densities, cell layer viability after 26 h growth, MTT incubation time, Ouabain and Veratridine treatment and solvent and matrix effects. A step-by-step protocol was defined identifying five viability controls for the validation of CBA-N2a results. Specific detection of two voltage gated sodium channel activators, pacific ciguatoxin (P-CTX3C) and brevetoxin (PbTx3) and two inhibitors, saxitoxin (STX) and decarbamoylsaxitoxin (dc-STX) was achieved, with EC_{50} values of 1.7 ± 0.35 pg/mL, 5.8 ± 0.9 ng/mL, 3 ± 0.5 ng/mL and 15.8 ± 3 ng/mL, respectively. When applied to the detection of ciguatoxin (CTX)-like toxicity in fish samples, limit of detection (LOD) and limit of quantification (LOQ) values were 0.031 ± 0.008 and 0.064 ± 0.016 ng P-CTX3C eq/g of flesh, respectively. Intra and inter-assays comparisons of viability controls, LOD, LOQ and toxicity in fish samples gave coefficients of variation (CVs) ranging from 3% to 29%. This improved test adaptable to either high throughput screening or composite toxicity estimation is a useful starting point for a standardization of the CBA-N2a in the field of marine toxin detection.

Keywords: CBA-N2a; standardization; matrix effects; absorbance data; ciguatoxins; brevetoxins; saxitoxins; biological sample; seafood safety

Key Contribution: The present work is a revisit of several parameters of the CBA-N2a that improved assay implementation and reliable toxin detection. This study resulted in a practical guide accessible to both supervised beginners and experienced users, provided in the Supplementary Materials. This revisited CBA-N2a allowed for the improved detection of toxins acting on the voltage gated sodium channels (ciguatoxin, brevetoxin and saxitoxin) as evidenced by CTX estimation data obtained from fish samples of known toxic status.

1. Introduction

The bio-accumulation of marine biotoxins produced by phytoplankton in filter-feeding invertebrates and finfish not only poses significant health threats to consumers but also has detrimental effects on the economies of nations highly dependent on seafood consumption for their subsistence [1–3]. Among these potent marine biotoxins are neurotoxins acting on the voltage gated sodium channels (VGSCs) of excitable cells, namely VGSC activators such as brevetoxins (PbTxs) and ciguatoxins (CTXs), and VGSC inhibitors such as saxitoxins (STXs) and tetrodotoxins (TTXs) [1,2]. Detection and quantification of these groups of marine toxins remain highly challenging due to the wide range of congeners present in trace amounts in contaminated biological matrices [4]. In addition, an official reference method is still lacking for most of these toxins due to the poor availability of certified reference standards [5]. Although the mouse bioassay has been recommended by the European Food Safety Authority (EFSA) for years until 2015, the European Union now strongly advocates the use of analytical techniques based on liquid chromatography coupled with mass spectrometry in tandem (LC-MS/MS) and high performance liquid chromatography with fluorescence detection (HPLC-FLD) for the detection of lipophilic and hydrophilic toxins, respectively [6–10]. However, the need for alternative high throughput methods for toxin screening and quantification purposes remains [8,11]. Among the readily available and most widely used in vitro methods, the functional assay known as the cell-based assay (CBA) that uses a neuroblastoma (N2a) cell line appears as the most promising one [2,12].

Based on the mode of action of VGSC toxins, the CBA-N2a was initially developed to detect two VGSC inhibitors, STXs and TTXs [13]. Those marine toxins have no cytotoxic effect on N2a cells and their detection requires the addition of the sodium/potassium (Na^+/K^+) ATPase pump blocker ouabain (O), together with the sodium-channel activator veratridine (V), which induces permanent activation of the VGSCs [14]. Under this O/V treatment (OV^+ conditions), sodium influx resulted in the cellular swelling and subsequent death of N2a cells [13]. Following the addition of STX or TTX, an increase of cell viability is observed as the two inhibitors counteract the effect of ouabain and veratridine. At that time, viability measure was achieved by a visual estimation of the morphological changes of N2a cells and the enumeration of viable cells under an inverted microscope. This assay performed in a 96-well microplate format has been applied to the detection of Paralytic Shellfish Poisoning (PSP) toxin standards [13] and to assess the presence of PSP toxins in biological extracts [15,16]. Standardization was also achieved using a certified STX standard for the detection of PSP toxins in shellfish and dinoflagellate cell extracts using automated endpoint determination with final readout based on crystal violet staining [17]. Further, this CBA-N2a was applied to detect VGSC inhibitors in marine bacterial supernatants using neutral red for final estimation of toxicity [18] and was subsequently extended to the detection of PbTxs and CTXs [19,20]. Contrary to VGSC inhibitors, these two VGSC activators induce a decrease in N2a cell viability in OV^+ conditions [19,20], whereas no cytotoxic effect is observed in the absence of O/V treatment [21]. These same authors also introduced a more workable measure of cell viability using the 3-(4,5-dimethyl-2-thiazol)-2,5-diphenyl-2H-tetrazolium bromide (methylthiazolyldiphenyl-tetrazolium bromide) colorimetric assay (known as MTT assay), as modified from previous protocols aiming at assessing either cellular growth and survival [22] or chemosensitivity in established cell lines [23]. The MTT assay is widely used to measure cell viability and proliferation [24–27], and is based on the capacity of mitochondrial dehydrogenase enzymes present in living cells to reduce tetrazolium salts into an insoluble purple formazan product [28] localized in lipids droplets [29]. The readout step was first established using dimethyl-sulfoxide (DMSO) to dissolve formazan products [23], the amount of the formazan produced is directly linked to the number of viable cells remaining at the end of the assay. This estimation is achieved by means of an automated readout. Two patents describing an improved and simplified CBA-N2a were eventually filed, leading to great expectations among the scientific community that this bioassay could soon be routinely used to detect both activators and inhibitors of sodium channels [30,31]. However, neither commercial kit nor detailed protocol is available to date.

Briefly, the CBA-N2a originally set by Manger et al. [19,20] consists of three major steps; (i) cell seeding of 100,000 cells/200 µL/well in a 96-well microplate left to grown for ≈ 24 h, (ii) O/V treatment followed by cell layer exposure to toxin standards and/or biological samples during 24 to 48 h, depending on the targeted toxin activity (final reaction volume of 230 µL/well) and (iii) N2a cell viability readout at 570 nm using the MTT colorimetric assay. Practically, the detection of VGSC activators and inhibitors by CBA-N2a requires 3 and 4 days, respectively.

Since 1993, the CBA-N2a has been widely used in a number of studies for the detection of VGSC toxins and related toxin families. However, a comprehensive review of the literature shows that the protocol initially defined by Manger et al. [19,20] has undergone numerous changes: for instance, (i) up to 13 different cell seeding densities have been tested, ranging from 10,000 to 250,000 cells/well [32–88] suggesting that cell confluence likely varied substantially between studies, especially when estimated by eye measurement; (ii) the culture medium established for cell layer implementation, which formerly used 10% of fetal bovine serum (FBS), was often reduced to 5% in many studies [32,33,36–39,42,46,48,54,59,69,70,74,77,78,81–83,85,86,89]; (iii) the confluence level reached by N2a cells after 24 h growth varied from non-confluent cells [66] to >90% confluence [48,59,63–65,72,73,77,78,83] or even was not specified in some studies [33–38,40–43,47,49–58,61,66,67,69,70,74–76,79–82,85,87–90].

Other steps of the CBA-N2a have also been the subject of substantial modifications. For instance, although many studies used a 500/50 μM (1:10 ratio) for O/V treatment [32–49,75,89,91,92] as initially defined by Manger et al. [19,20,30,31], numerous changes in O/V concentrations [54–57,59,61,63–65,71–75,77–83,85,87,93–95], O/V ratios [50–52,69,86] and reaction volumes [32,33,35,36,49,51,53,59,61,63,64,69,73,75,77–80,84,86–90] were further proposed. Interestingly, in many of these studies, the effects on cell viability resulting from all these modifications were rarely addressed but, when reported, showed incongruent results. For instance, O/V treatment at 500/50 μM was shown to induce highly variable effects between 20–92% [34,45,68,96]. Conversely, similar effects (10–56%) were reportedly obtained at lower O/V concentrations [50,86]. The most significant modification was the use of two different O/V treatments, i.e., 100/10 and 300/30 μM, in order to induce 20% and 80% of cell mortality, respectively, for a more reliable assessment of the effects of VGSC toxin activators and inhibitors on N2a cells [83]. Several changes were also made to the final reaction volume (230 μL) at the last step of the CBA-N2a, i.e., cell exposure to toxin standards or biological samples, with volumes varying from 100 to 210 μL [32–36,45,49,51,53,59,61,63,64,69,73,75,77–80,84,86–90,95]. Moreover, measuring cell viability by means of the MTT colorimetric assay is classically performed at 570 nm, however, different wavelengths were also tested that ranged from 490 to 595 nm [33,35,36,38–42,45,50,56,57,61,64,67–69,71,75,76,79,80,84,85,87,88,90,93,97]. Finally, water-soluble tetrazolium reagents were sometimes preferred to the former water insoluble formazan product, with the use of WST-8 [49,60] or commercial kits such as CellTiter 96® Aqueous One solution [33,38,69,86], Cell Counting Kit-8 [47] or XTT Cell Proliferation Assay kit [98].

This wide range of "in-house" methods described in the literature highlights the present lack of a consensus, standardized protocol for CBA-N2a, making any attempt to compare CBA data between assays and/or laboratories difficult or impossible. It also greatly hampers all current efforts to establish the CBA-N2a tool as a potential alternative reference method to LC-MS/MS. In this context, the present work aims at a comprehensive revisit of the CBA-N2a by evaluating the effects of six key parameters that are critical in obtaining reliable toxicity results, i.e., cell seeding densities, cell layer viability after 26 h growth, MTT incubation time, O/V treatment and solvent and matrix effects. To evaluate its robustness, the newly improved protocol thus established was further applied to the detection of toxins active on VGSC families (activators vs. inhibitors), and the estimation of CTX-like toxicity in fish samples of known ciguatoxic status.

2. Results

2.1. *Characterization of N2a Cell Growth and Initial Viability*

The N2a CLL 131 cell line used in this study displayed a typical growth curve characterized by a short lag phase (slow cell growth), followed by a log phase (exponential proliferation of cells and consumption of nutrients of the culture medium) until a maximum density of ≈ 100,000 cells/well is reached, and a late stationary/senescence phase (reduced cell proliferation) (Figure 1).

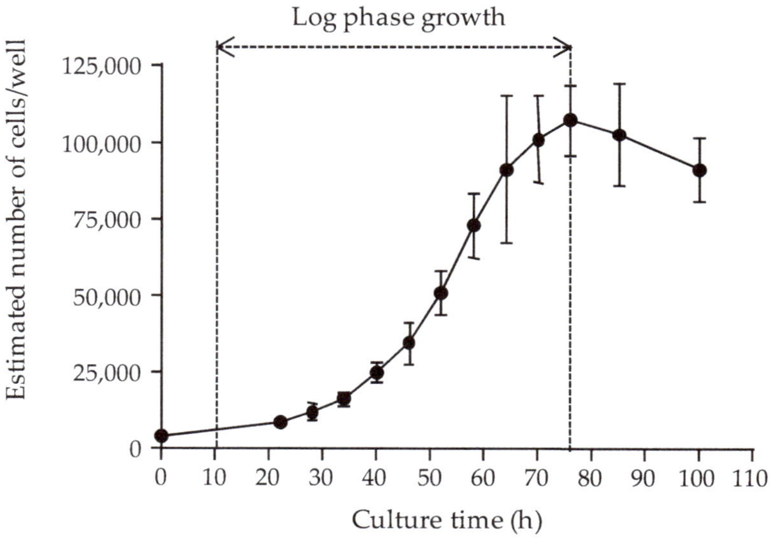

Figure 1. Characteristic growth pattern of the neuroblastoma (N2a) cell line used in this study. Densities in 96-well microplates were extrapolated from growth experiments conducted over a 4 days period in 25 cm^2 culture flasks using a 10% fetal bovine serum (FBS) culture medium. Data represent the mean ± standard deviation (SD) of one experiment (N2a cells at 383 passages (P)), with $n = 10$ counts for each point. Coefficients of variation (CV) ranged from 10.8% to 26.3%.

The linearization of the log phase of the growth curve was defined by the following equation:

$$Y = 0.0578X + 7.7948 \ (r^2 = 0.9974) \quad (1)$$

in which X is the culture time (hours) and Y is the Ln-transformed cell number. Based on this equation, it was concluded that the N2a growth curve was characterized by a 9.8 h lag phase (Y = Ln (4288)) and that the cell number increased by two-fold after an additional 12 h. Moreover, a cell seeding density of 50,000 ± 10,000 cells/well allowed reaching a maximum cell density of ≈ 100,000 cells/well after 22 h culture time. For more convenience, a culture time of 26 h post-seeding was selected in all further experiments.

Next, the N2a cell initial viability after 26 h growth was compared at ten different cell seeding densities ranging from 10,000 to 100,000 cells/well and in two culture conditions (5% and 10% FBS growth medium) (Figure 2). Results showed that (i) the highest absorbance value is consistently obtained at a cell seeding density around the maximum number of cells supported by microplate wells, and (ii) absorbance values increase in proportion with MTT incubation times (Figure 2). For instance, the maximum absorbance value measured after 26 h culture time and 45 min MTT incubation time was 1.4 while selecting a cell seeding density of 50,000 ± 10,000 cells/well (vertical dotted lines, Figure 2) allowed to reach absorbance values comprised between 1 and 1.25 (horizontal dotted lines, Figure 2). This absorbance range was considered optimal when the detection of a decrease in cell viability is sought. Finally, no differences were observed when using either 5% or 10% FBS growth medium, which indicates possible saving opportunities on this expensive reagent at this step of the CBA-N2a

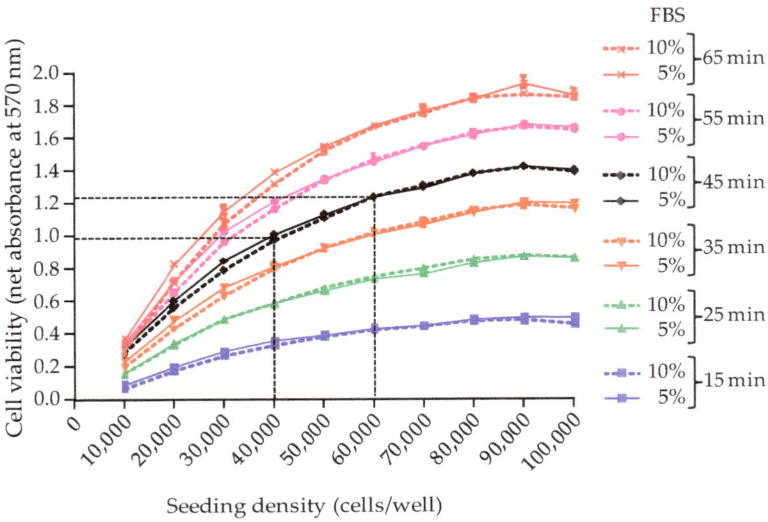

Figure 2. Initial viability of N2a cells observed in 96-well microplates after 26 h growth in 5% FBS (full line) and 10% FBS (dotted line) culture medium, at different cell seeding density. Six distinct MTT incubation times were also tested: 15 min (blue); 25 min (green); 35 min (orange); 45 min (black); 55 min (pink); 65 min (red). Data represent the mean ± SD of one microplate (N2a cells at 535 P), each point tested in six wells. Mean CVs were <3%. Absorbance values were measured at 570 nm via the MTT assay.

Based on these results, all further CBA-N2a experiments were conducted as follows:

- Implement cell layer in microplates using a cell seeding density of 50,000 ± 10,000 cells/well in 200 µL of a 5% FBS culture medium, in order to reach a maximum cell density of 100,000 ± 20,000 cells/well after ≥22 h of culture.
- Conduct the MTT assay at an incubation time of 45 min, in order to reach an absorbance value ≥1.0 that is used to define N2a "initial viability".
- For each experiment, dedicate a separate microplate to measure the N2a cell initial viability after 26 h of growth defined as the "Reference Cell Viability" control (RCV control).

2.2. *Characterization of N2a Cell Final Viability*

The second step of CBA-N2a is the exposure of N2a cells to VGSC activators or inhibitors, in OV^- or OV^+ conditions. Following an additional culture time of 19 h overnight, the final viability of N2a cells was assessed as previously described.

2.2.1. N2a Cell Final Viability in OV^- Conditions

A final cell viability lower than the initial cell viability (as measured in the RCV control) was observed only with 1% FBS growth medium at cell seeding density > 40,000 cells/well (Figure 3). Conversely, all other growth media allowed to reach a final viability higher than the one displayed by the RCV control, regardless of the cell seeding density (Figure 3), suggesting a complete renewal of the culture medium with 2% FBS growth medium is sufficient to ensure a stable cell viability during 19 h additional culture of the toxin exposure. A better stability of cell viability was consistently achieved at cell seeding densities > 40,000 cells/well. In other words, working at a cell seeding density of 50,000 ± 10,000 cells/well will allow to reach an absorbance value ≥1, close to the one measured in the RCV control (Figure 3).

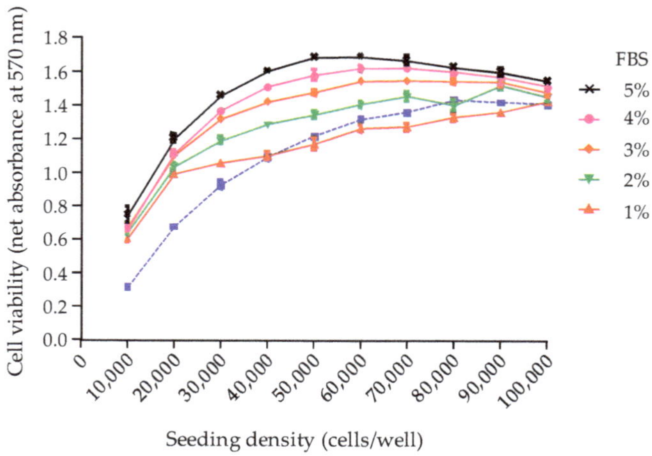

Figure 3. Final viability of N2a cells observed at different seeding densities when cultured 26 h in 5% FBS culture medium followed by 19 h in fresh growth medium supplemented with 1% FBS (red), 2% FBS (green), 3% FBS (orange), 4% FBS (pink) and 5% FBS (black). The initial viability measured in the Reference Cell Viability (RCV) control microplate after 26 h of growth is represented by the blue dotted curve. Data represent the mean ± SD of one microplate (N2a cells at 537 P), each point tested in six wells. Absorbance values were measured at 570 nm via the MTT assay, after a 45 min MTT incubation time.

Based on these results, the following procedure was selected to ensure that, at this step of the CBA-N2a, a final cell viability close to the initial cell viability is obtained in OV^- conditions:

- After cell layer implementation (26 h after seeding), remove the used culture medium from microplates (except for the RCV control plate which is sacrificed to get the initial N2a cell viability via the MTT assay).
- Replace with 200 µL of fresh 2% FBS growth medium prior to an additional culture time of 19 h.
- Measure the absorbance data in control wells (n = 3) in the absence of O/V treatment (COV^-) to characterize the N2a cell final viability in OV^- conditions.
- Compare absorbance data in COV^- control vs. RCV control to verify cell viability is maintained in OV^- conditions.

2.2.2. N2a cell Final Viability in OV^+ Conditions

As expected, absorbance data measured at 0/0 µM (OV^- conditions) was close to the one of the RCV control (Figure 4). At O/V concentrations ranging from 10/1 to 80/8 µM, a "protective effect" was observed as evidenced by an increase in cell viability of ≈ 20% above the RCV control (Figure 4). Between 80/8 and ≈ 110/11 µM, although a slight decrease in absorbance data was observed, these values were consistently found above the RCV control (Figure 4), suggesting a slightly toxic effect, although regarded as "non-destructive effect" of O/V treatment on N2a cells (Figure 4). At higher concentrations, however, cell viability decreased in a dose-dependent manner down to the level observed for DMSO control, which was indicative of a "destructive effect" of O/V treatment on N2a cells (Figure 4). Above 300/30 µM, the effect of O/V on cell viability was considered "lethal" with the complete elimination of cell viability (Figure 4). Concerning the variability of O/V treatment obtained across a wide range of N2a cell passages from 384 to 804 P, the CVs were below 12.4% for the "non-destructive" O/V treatment between 80/8 and 100/10 µM selected for the detection of VGSC activators. When seeking to detect VGSC inhibitors, the absorbance values were close to 0 showing non significative CVs above 45% for the "destructive" O/V treatment between 270/27 and 300/30 µM.

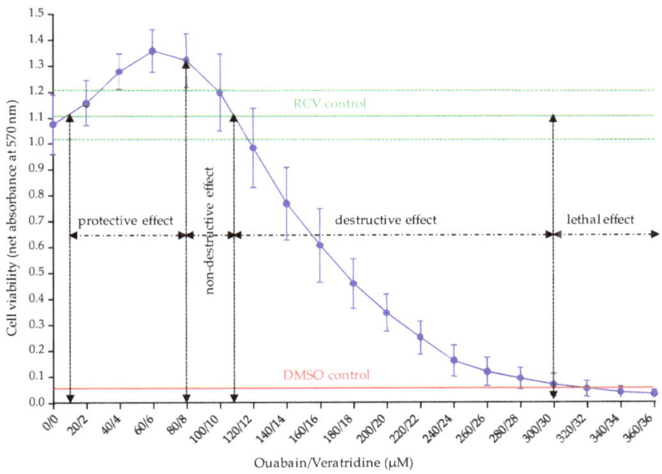

Figure 4. Dose-response curve of N2a cells when exposed to increasing concentrations of Ouabain and Veratridine (O/V) treatments ranging from 0/0 to 360/36 μM. Data represent the mean ± SD of one microplate in six independent experiments corresponding to cell passage numbers of 384, 542, 800, 801, 803 and 804 P, each point run in triplicate. The mean absorbance ± SD values corresponding to the RCV control (horizontal green line and dotted green lines) and DMSO control (horizontal red line) were determined at 1.105 ± 0.096 and 0.042 ± 0.001, respectively. Absorbance values were measured at 570 nm via the MTT assay, after a 45 min MTT incubation time. The four different effects induced by increasing concentrations of O/V treatment on N2a cells, i.e., protective, non-destructive, destructive and lethal are also represented.

Based on these results, the O/V treatment conditions best adapted to the type of activity we seek to detect on target cells (e.g., activation or inhibition of VGSCs on N2a cells) were defined as follows:

- Select non-destructive O/V treatment conditions (between 80/8 and 100/10 μM) in order to induce a final cell viability slightly above the one observed in the RCV control, so that any loss in cell viability can be assigned to the specific activity of VGSCs activators. In the following experiments (Sections 2.4.2 and 2.4.3), 100/10 and 85.7/8.57 μM (final concentrations) were used.
- Select destructive O/V treatment conditions close to the lethal effect (between 270/27 and 300/30 μM) in order to induce a final cell viability slightly above the one observed in the DMSO control, so that any increase in cell viability can be assigned to the specific activity of VGSC inhibitors. In the following experiments (Section 2.4.2), 270/27 μM (final concentrations) were used.
- Measure the absorbance data in control wells ($n = 3$) in the presence of O/V treatment (COV^+) to characterize the N2a cell final viability in OV^+ conditions.
- Compare absorbance data in COV^+ control vs. RCV control to verify the efficiency of the O/V treatment conditions applied in the experiment.

2.3. *Characterization of the Unspecific Effects of Solvent and Dry Extract on N2a Cell Viability*

The issue of potential solvent toxicity and sample extract matrix effects on N2a cell viability was also taken into consideration.

2.3.1. Solvent Effects

Using methanol (MeOH), a progressive loss in N2a cell viability is observed in 100/10 μM OV^+ conditions at solvent concentrations ≥0.8%, giving a mean absorbance value ≈ 25% lower than the one measured in the RCV control at the highest concentration tested (Figure 5a). In OV^- conditions, an opposite effect is observed as evidenced by an increase in cell viability also at concentrations ≥0.8% (Figure 5a). Conversely, regardless of the OV conditions, DMSO consistently induced an abrupt decrease in cell viability at concentrations ≥1% until the complete elimination of viable N2a cells (Figure 5b). These findings indicate that both solvents can be used provided a concentration <0.8% is selected and that the maximum concentration of solvent can start at 0.5%. Practically, the use of non-volatile DMSO was preferred for the preparation of stock solutions (especially for toxin standards) as it is non-hazardous and likely to ensure stable toxin concentration over long periods of storage.

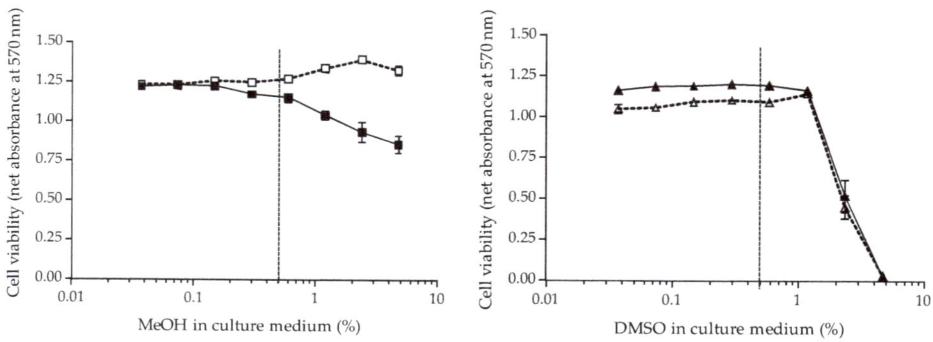

(**a**) (**b**)

Figure 5. Dose-response curves of N2a cells when exposed to increasing concentrations of two solvents, in OV^- (open symbols) and OV^+ (solid symbols) conditions at 100/10 µM (final concentrations); (**a**) MeOH (□/■) and (**b**) DMSO (Δ/▲) were tested using N2a cells at 551 and 549 P, respectively. Data represent the mean ± SD of one microplate, with each point run in triplicate. Absorbance values were measured at 570 nm via the MTT assay, after a 45 min MTT incubation time. The initial cell viability in the RCV control was 1.139 ± 0.021 and 0.995 ± 0.031 and the final cell viability in the absence of O/V treatment (COV^-) control was 1.217 ± 0.025 and 1.047 ± 0.023 for (**a**,**b**), respectively. The dotted vertical line corresponds to the maximum solvent concentration 0.5% for solvent interferences.

Consequently, a dilution of at least 1:200 giving the first final concentration (C1) of the toxin standard/sample extract stock solutions in MeOH or DMSO was achieved in all further CBA-N2a experiments, as follows:

- Initial dilution of at least 1:10 in 2% FBS culture medium,
- Final dilution at 1:21 by direct addition of 10 µL of initial dilution in 200 µL of growth medium. This will ensure the final solvent concentration in wells does not exceed 0.5%.

2.3.2. Biological Matrix effects

In this study, the dry extract weights (DEW) obtained from 10 g of fish flesh ranged from 2.7 to 4.3 mg (Section 5.2), but they have been shown to sometimes vary up to ten-fold in other studies (data not shown). When tested at similar concentrations in OV^- conditions, all extracts induced a final cell viability slightly above the one observed in the RCV control at concentrations ≥1500 pg/µL, whereas cell viability could increase up to 70% as observed for sample Emer05 at 10,000 pg/µL (or 30.08 ± 5.92 µg fish flesh equivalent/µL) (Figure 6). Above this concentration, a decrease in cell viability rapidly occurred, suggesting an unspecific cytotoxicity on N2a cells likely due to matrix effects (Figure 6).

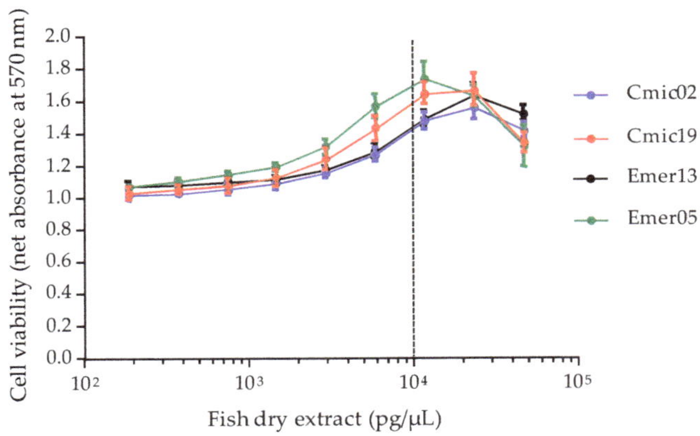

Figure 6. Dose-response curves of N2a cells in OV⁻ conditions when exposed to increasing concentrations of LF90/10 dry extracts prepared from *Chlorurus microrhinos* Cmic02 (blue), Cmic19 (red), *Epinephelus merra* Emer13 (black) and Emer05 (green) fish samples. Data represent the mean ± SD of three independent microplates (N2a cells at 800, 801 and 803 P), each point run in triplicate. Absorbance values were measured at 570 nm via the MTT assay, after a 45 min MTT incubation time. The cell viability in the RCV control was determined at 1.143 ± 0.009, while the final cell viability in COV⁻ control was 1.037 ± 0.053. The dotted vertical line corresponds to the maximum concentration of dry extract (MCE = 10,000 pg/µL) for matrix interferences.

Based on these results, the MCE for LF90/10 fractions was set at 10,000 pg/µL for this study in all further experiments. Prior to CBA-N2a toxicity analysis, dry extract solutions of fish samples were thus prepared as follows:

- For all fish samples, resuspend LF90/10 dry extract in MeOH or DMSO at 10 mg/mL, then proceed with an intermediate 1:50 dilution of this previous stock solution.
- Add 10 µL of the solution thus obtained in 200 µL of growth medium to test the highest final concentration below the MCE, i.e., 9524 pg/µL, in the absence vs. under non-destructive O/V treatment.

This revisited CBA-N2a resulted in a practical guide that is presented in the Supplementary Materials.

2.4. Application to the Detection of VGSC Activators and Inhibitors

2.4.1. N2a Cell Initial Viability

To test the relevance of the six parameters revisited in Sections 2.1–2.3, this improved protocol was further applied to the detection of different toxin standards acting on VGSCs.

First, the initial viability of N2a cells prior to toxin exposure was assessed by measuring absorbance values in RCV control using the MTT colorimetric assay in three independent experiments. The metabolization of MTT into formazan by N2a viable cells resulted in a uniform blue color visible in all wells, proof that a high cell layer confluence was attained. Following the addition of DMSO and cell lysis, a dark purple color was released in each well (Figure S1).

The net absorbance data obtained for RCV controls in all three experiments were indicative of a high viability of N2a cells (mean value between 1 and 1.25) and were also highly reproducible, with CVs below 5% (Table 1).

Table 1. Initial viability of N2a cells as assessed in RCV controls (cell layer ≈ 100,000 cell/well obtained after 26 h of growth).

Variability	Experimental Plates	Absorbance Raw Data (1)	DMSO Control (2)	RCV Control (1)–(2)	CVs (%)
Inter-assay *	3	1.170 ± 0.050	0.043 ± 0.0001	1.127 ± 0.050	4.4

* Data represent the mean ± SD of one microplate in three independent experiments (N2a cells at 662, 663 and 666 P).

2.4.2. Detection of VGSC Activators and Inhibitors

In OV^- conditions (upper half of the microplates), the same dark purple color was observed for COV^- control and all concentrations tested regardless of the toxin standard (Figure S1). Conversely, in OV^+ conditions (bottom half of the microplate), a progressive fading of this dark purple color was observed with increasing concentrations of P-CTX3C and PbTx3, until complete discoloration, signing the total death of N2a cells (Figure S1). Otherwise, a progressive darkening of the pale color observed at low concentrations occurred with increasing concentrations of STX and dc-STX, signing a progressive restoration of N2a cell viability (Figure S1).

For VGSC activators, net absorbance values showed that COV^- and COV^+ control absorbance data were close to that of the RVC control. The same observation applies to VGSC inhibitors except in OV^+ conditions where absorbance data of COV^+ controls were in the range of 0.1.

Inter-assay comparison in three independent experiments showed that net absorbance data are reproducible with CVs of 6.1% and 5.4% for COV^- controls, and 1.6% and 4.5% for COV^+ controls of P-CTX3C and PbTx3, respectively. Likewise, net absorbance data of COV^- controls showed CVs of 6% and 3.9% of STX and dc-STX, respectively, but 41.4% and 37.5% for COV^+ controls, these latter values being close to 0 absorbance.

In the presence of the two VGSC activators, net absorbance data of all nine concentrations obtained in OV^- conditions remained close to the ones measured in COV^-, COV^+ and RCV controls regardless of the toxin concentrations tested, whereas in OV^+ conditions, a sigmoidal dose-response curve was obtained for both P-CTX3C and PbTx3 (Figure 7a,b). In the presence of the two VGSC inhibitors, net absorbance data of all nine concentrations obtained in OV^- conditions remained close to the ones measured in COV^- and RCV controls regardless of the toxin concentrations tested, whereas in OV^+ conditions, a sigmoidal dose-response curve was obtained for both STX and dc-STX (Figure 7c,d).

Table 2 details several characteristic parameters of CBA-N2a curves. The EC_{50} values roughly corresponded to twice that of EC_{80} for VGSC activators and to 2.5-fold higher than that of EC_{20} for VGSC inhibitors. Overall, high reproducibility was found for top and bottom absorbances as well as negative Hillslopes with CVs < 5% for P-CTX3C and PbTx3, and CVs < 10% for STX and dc-STX (Table 2). Regarding VGSC activators, CVs were ≈ 19% and between 17% and 21% for EC_{80} and EC_{50} values, respectively. For VGSC inhibitors, CVs of EC_{20} and EC_{50} varied between 10.3% and 17.5% for STX, and between 19.2% and 26.3% for dc-STX. Comparing the potency within VGSC acting toxin family, P-CTX3C was approximately 3300-fold more potent than PbTx3 and STX was approximately 4-fold more potent than dc-STX (Table 2).

All these findings showed that this newly improved CBA-N2a allowed for specific and sensitive detection of VGSC activators and inhibitors.

Based on these results, quality check (QC) controls were defined as follows:

- The "$QCOV^-$" is the quality control that serves to check for the final viability of N2a cells after 19h of additional culture in the absence of O/V treatment and in the presence of VGSC acting toxins.
- The "$QCOV^+$" is the quality control that serves to check for the final viability of N2a cells after 19h of additional culture under O/V treatment and in the presence of VGSC acting toxins.

The QC controls are used to verify the specific action of VGSC acting toxins (Figure S2) and their concentration is selected at the EC_{50} of a given standard toxin.

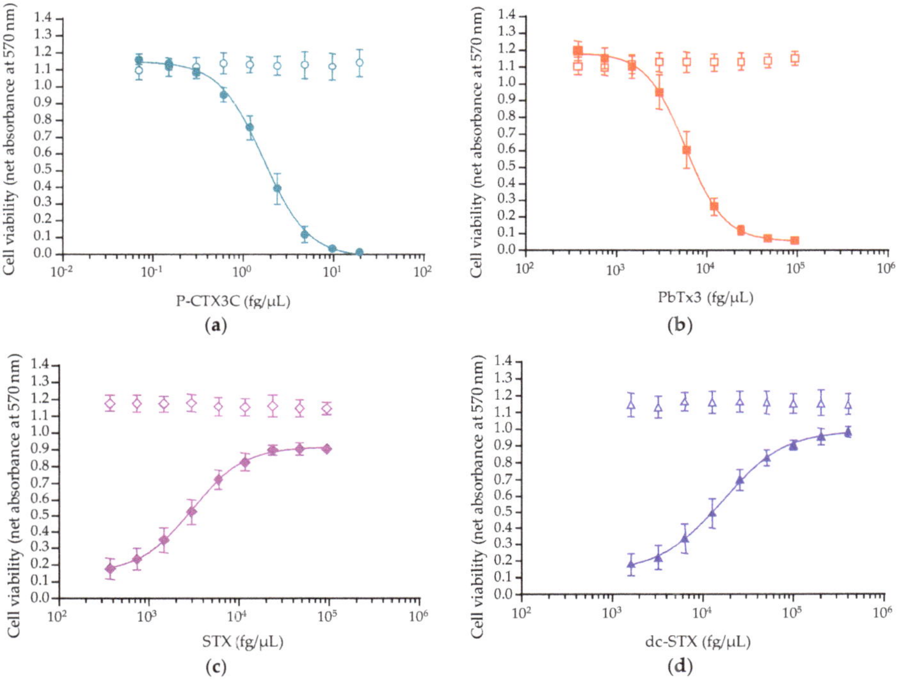

Figure 7. Dose-response curves displayed by N2a cells when exposed to nine increasing concentrations of toxin standards after 19 h exposure time, in OV^- (open symbols) and OV^+ (solid symbols) conditions at 100/10 μM for voltage gated sodium channels (VGSC) activators and 270/27 μM for VGSC inhibitors (plain symbols): (**a**) P-CTX3C (○/●); (**b**) PbTx3 (□/■); (**c**) (STX) (◊/♦); (**d**) dc-STX (Δ/▲). Data represent the mean ± SD of one microplate in three independent experiments (N2a cells at 662, 663 and 666 P), with each concentration run in triplicate. Net absorbance values were 1.101 ± 0.067 and 1.077 ± 0.058 in COV^- controls (n = 3), and 1.209 ± 0.014 and 1.226 ± 0.055 in COV^+ controls (n = 3) for (**a**,**b**), respectively, for VGSC activators. Net absorbance values were 1.170 ± 0.046 and 1.113 ± 0.067 in COV^- controls (n = 3), and 0.136 ± 0.051 and 0.145 ± 0.060 in COV^+ controls (n = 3) for (**c**,**d**), respectively.

Table 2. Dose-response curve parameters of cell-based assay (CBA)-N2a when detecting VGSC activators or inhibitors.

Parameters	P-CTX3C **	PbTx3 **	STX **	dc-STX **
Top absorbance	1.149 ± 0.028	1.182 ± 0.060	0.917 ± 0.017	0.990 ± 0.038
Bottom absorbance	−0.017 ± 0.028	0.057 ± 0.013	0.149 ± 0.073	0.130 ± 0.049
Hillslope	−1.802 ± 0.074	−2.052 ± 0.102	1.545 ± 0.127	1.290 ± 0.126
EC_{80} * (fg/μL)	0.787 ± 0.155	2950 ± 537	ND	ND
EC_{50} * (fg/μL)	1.700 ± 0.354	5800 ± 926	2982 ± 523	15851 ± 3050
EC_{20} * (fg/μL)	ND	ND	1195 ± 123	5456 ± 1434

* Data represent the mean ± SD of one microplate in three independent experiments (N2a cells at 662, 663 and 666 P).
** OV^+ conditions: 100/10 and 270/27 μM final concentrations for VGSC activators and inhibitors, respectively.
ND = not determined.

2.4.3. Composite Toxicity Estimation of VGSC Activators in Fish Samples

Four fish samples were analyzed: two steephead parrot-fish (*Chlorurus microrhinos*, Scaridae, Cmic02 and Cmic19) and two honeycomb groupers (*Epinephelus merra*, Serranidae, Emer05 and Emer13). To evaluate the composite CTX-like toxicity, a set of standards must be run in parallel with samples in order to calibrate an experiment properly. For this purpose, P-CTX3C standard was tested in parallel with fish matrix, in the absence vs. under non-destructive O/V treatment at 85.7/8.57 µM (final concentrations), as the EC_{50} of standards are further needed to infer toxin content in biological samples.

A high repeatability and reproducibility of viability data were obtained for RCV, COV^-, COV^+ controls as well as $QCOV^-$ (Table 3), with CVs < 7%, whereas a lower reproducibility was observed for $QCOV^+$ absorbance data with a CV of 17.3% inherent to the cytotoxicity of PbTx3 at EC_{50} (Section 2.4.2).

Table 3. Assessment of five viability controls useful to validate the detection of VGSC activators in fish matrix using the revisited CBA-N2a.

Variability	RCV Control	COV^-	COV^+	$QCOV^-$	$QCOV^+$
Intra-assay *	1.122 ± 0.046	1.264 ± 0.068	1.130 ± 0.029	1.257 ± 0.067	0.448 ± 0.038
Inter-assay **	1.132 ± 0.052	1.165 ± 0.074	1.129 ± 0.078	1.174 ± 0.074	0.409 ± 0.071

* Data represent the mean ± SD of three microplates in one experiment (N2a cells at 810 P). ** Data represent the mean ± SD of one microplate in three independent experiments (N2a cells at 795, 797 and 798 P).

Moreover, statistical analyses by means of the Wilcoxon test showed no significant difference between RCV control and COV^+, and between $QCOV^-$ and COV^- for both intra- and inter-assay (Table S1). Conversely, significant differences were observed between $QCOV^+$ and all other controls. In addition, the comparison between intra- and inter-assay values showed no significant differences for COV^+ values, nor between RCV control and COV^+ values showing that final viability was similar to initial viability under non destructive O/V treatment (Table S1).

The dose-response curves thus obtained for P-CTX3C (data not shown) allowed to establish another set of EC_{80} and EC_{50} (Table 4) which were compared to those of Table 2 (Section 2.4.2). Statistical comparison by means of the Wilcoxon test showed no significant differences between EC_{80} and EC_{50} values obtained under two distinct O/V treatments, i.e., 100/10 and 85.7/8.57 µM with p-values > 0.4 (Table S2).

Table 4. Variabilities of EC_{80} and EC_{50} for two VGSC activators (P-CTX3C and PbTx3) using the revisited CBA-N2a as assessed under non-destructive O/V treatment conditions.

Variability	O/V Treatment	P-CTX3C		PbTx3	
		EC_{80} (fg/µL)	EC_{50} (fg/µL)	EC_{80} (fg/µL)	EC_{50} (fg/µL)
Inter-assay *	100/10 µM	0.787 ± 0.155	1.700 ± 0.354	2950 ± 538	5800 ± 929
Intra-assay **	85.7/8.57 µM	0.964 ± 0.055	2.004 ± 0.190	3230 ± 270	6241 ± 388
Inter-assay ***	85.7/8.57 µM	0.825 ± 0.236	1.730 ± 0.414	3015 ± 347	5946 ± 788

* Data represent the mean ± SD of one microplate in three independent experiments (N2a cells at 662, 663 and 666 P). ** Data represent the mean ± SD of three microplates in one experiment (N2a cells at 810 P). *** Data represent the mean ± SD of one microplate in three independent experiments (N2a cells at 795, 797 and 798 P).

Further, the limit of detection (LOD) and the limit of quantification (LOQ) of the CTX-like toxicity in fish were determined using the MCE of 10,000 pg dry extract/µL and EC_{80} and EC_{50} values obtained for P-CTX3C exclusively run in parallel with fish samples under 85.7/8.57 µM OV treatment. The LOD and LOQ showed high repeatability with CVs < 9.5% (intra-assay), whereas a lower reproducibility was noticed with CVs > 23.9% (Table 5). The Wilcoxon test also confirmed no significant differences between intra- and inter-assay data with p-values of 0.7 and 0.4 for LOD and LOQ, respectively. Hence, a mean LOD value was established at 0.089 ± 0.017 ng/mg of dry extract for P-CTX3C, with LOQ values about twice that of LOD.

Table 5. Estimation of limit of detection (LOD) and limit of quantification (LOQ) of the CTX-like toxicity in fish flesh using the revisited CBA-N2a under non-destructive O/V treatment at 85.7/8.75 μM.

Variability	Sample ID	ng P-CTX3C eq/mg Dry Extract		ng P-CTX3C eq/g Fish Flesh	
		LOD	LOQ	LOD	LOQ
Intra-assay **	Cmic02	0.096 ± 0.006	0.200 ± 0.019	0.035 ± 0.002	0.072 ± 0.007
	Cmic19			0.041 ± 0.002	0.086 ± 0.008
	Emer05			0.030 ± 0.002	0.062 ± 0.006
	Emer13			0.026 ± 0.001	0.054 ± 0.005
Inter-assay *	Cmic02	0.083 ± 0.024	0.173 ± 0.041	0.030 ± 0.008	0.062 ± 0.015
	Cmic19			0.035 ± 0.010	0.074 ± 0.018
	Emer05			0.026 ± 0.007	0.054 ± 0.013
	Emer13			0.022 ± 0.006	0.047 ± 0.011

* Data represent the mean ± SD of one microplate in three independent experiments (N2a cells at 795, 797 and 798 P).
** Data represent the mean ± SD of three microplates in one experiment (N2a cells at 810 P).

Based on the "dry extract weight/fresh weight" (DEW/FW) ratio characterizing each fish sample (Section 5.2), the LOD and LOQ of P-CTX3C in fish flesh were determined and expressed in ng of P-CTX3C eq/g of fish flesh (Table 5) for further comparison with the EFSA and US Food and Drug Administration (FDA) advisory level. Although LOD and LOQ of the CTX-like toxicity varied from one fish to another (Table 5) due to differences in DEW/FW ratios, the Wilcoxon test showed no significant differences between intra- and inter-assay data regardless of the fish sample and P-CTX3C used as reference (Table S3). Hence, the mean LOD values were established at 0.031 ± 0.008 (CV = 25.4%) and LOQ = 0.064 ± 0.016 (CV = 24.3%) ng P-CTX-3C eq/g of fish flesh.

Application of the CBA-N2a to the detection of VGSC activators in four fish samples gave two distinct patterns as observed for VGSC activators (Figure S1). When exposed to increasing concentrations of fish extracts (Figure 8a,b), no cytotoxic effects were observed in OV^- and OV^+ conditions for Cmic02 and Emer13. Conversely, sigmoidal dose-response curves with a negative slope were obtained for Cmic19 and Emer05 fish samples in OV^+ conditions (Figure 8a,b), whose ciguatoxicity was previously characterized by fluorescent receptor binding assay (fRBA) and/or LC-MS/MS analyses (Section 5.1.3).

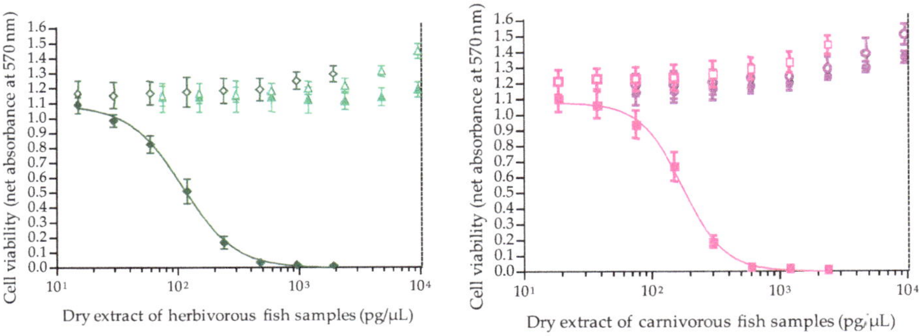

(a) (b)

Figure 8. Dose-response curves displayed by N2a cells when exposed to increasing concentrations of fish dry extracts (LF90/10) of two herbivorous fishes (**a**) and two carnivorous fishes (**b**) after 19 h exposure time in OV^- (open symbols) and OV^+ conditions 85.7/8.57 μM (solid symbols): Cmic02 (Δ/▲), Cmic19 (◊/♦), Emer13 (○/•) and Emer05 (□/■). Data represent the mean ± SD of one microplate in three independent experiments (N2a cells at 795, 797 and 798 P), with each concentration run in triplicate. The dotted vertical line corresponds to the maximum concentration of dry extract (MCE = 10,000 pg/μL) for matrix interferences.

The sigmoidal dose-response curves thus obtained for Cmic19 and Emer05 were fitted according to the four parameter logistic regression (4PL) model showing high repeatability and reproducibility for top absorbance and Hillslope, with CVs < 10% (Table 6). Absorbance values close to 0 were also consistently obtained at the highest concentration of fish extract tested (Table 6).

The mean EC_{50} and EC_{80} values determined in pg of dry extract/μL for Cmic19 and Emer05 also showed a high repeatability with CVs between 5% and 11.2%, respectively, whereas a lower reproducibility of these values was noticed with CVs between 12% and 21.6%, respectively (Table 6). The Wilcoxon test also confirmed no significant differences between intra- and inter-assay data with p-values of 0.7 and 0.2 for Emer05 and Cmic19, respectively. Hence, mean EC_{80} values were established at 63.6 ± 12.4 and 108.5 ± 13.7 pg/μL for Cmic19 and Emer05, respectively, with mean EC_{50} values 1.6 and 1.9 higher than that of EC_{80} for Emer05 and Cmic19, respectively. Based on EC_{50} values, Cmic19 dry extract was 1.5-fold more potent than Emer05 dry extract (Table 6).

Table 6. Dose-response curves parameters of CBA-N2a when detecting VGSC activators in fish samples.

Variability	Sample ID	Top Absorbance	Bottom Absorbance	EC_{80} *** (pg/μL)	EC_{50} *** (pg/μL)	Hillslope
Intra-assay *	Cmic19	1.064 ± 0.007	0.001 ± 0.004	71.2 ± 8	130.6 ± 11	−2.282 ± 0.132
	Emer05	1.110 ± 0.014	0.003 ± 0.004	114.7 ± 10	184.4 ± 10	−2.927 ± 0.271
Inter-assay **	Cmic19	1.084 ± 0.062	−0.003 ± 0.002	55.9 ± 12.1	109.5 ± 21.1	−2.056 ± 0.073
	Emer05	1.081 ± 0.101	0.000 ± 0.002	102.2 ± 15.9	171.3 ± 20.5	−2.675 ± 0.196

* Data represent the mean ± SD of three microplates in one experiment (N2a cells at 810 P). ** Data represent the mean ± SD of one microplate in three independent experiments (N2a cells at 795, 797 and 798 P). *** OV^+ conditions: 85.7/8.57 μM final concentrations.

In addition, the toxin contents in fish flesh were also estimated using the EC_{50} values determined for toxin standard and fish extract (Table 7).

Overall, CTX-like composite toxicity data showed high repeatability and reproducibility with CVs < 13% (Table 7). The Wilcoxon test showed no significant differences between intra- and inter-assays for toxin contents with p-values of 0.3 and 0.6 for Emer05 and Cmic19, respectively. Hence, the mean toxin content in Cmic19 and Emer05 was estimated at 6.66 ± 0.68 and 3.31 ± 0.35 ng P-CTX-3C eq/g fish flesh, respectively, indicating Cmic19 was twice as toxic as Emer05.

Table 7. CTX-like composite toxicity estimation in two ciguatoxic fish samples using the revisited CBA-N2a.

Variability	Sample ID	ng P-CTX3C eq/g Fish Flesh	
		Mean ± SD	CVs (%)
Intra-assay *	Cmic19	6.63 ± 0.74	11.2
	Emer05	3.37 ± 0.32	9.5
Inter-assay **	Cmic19	6.76 ± 0.57	8.4
	Emer05	3.10 ± 0.40	13

* Data represent the mean ± SD of three microplates in one experiment (N2a cells at 810 P). ** Data represent the mean ± SD of one microplate in three independent experiments (N2a cells at 795, 797 and 798 P).

3. Discussion

Having sensitive and specific detection methods available to protect consumers against poisoning risks due to seafood contaminated with marine biotoxins is a key component of food security monitoring programs worldwide [1,5]. However, in the absence of duly validated reference methods, many marine biotoxins are still unregulated [5]. This is the case for instance for CTXs, neurotoxins active on VGSCs that are implicated in ciguatera poisoning [5]. Among the many detection tests currently available for the detection of this group of compounds and, more widely, of toxins acting on voltage-gated sodium channels [8,99], the neuroblastoma cell-based assay or CBA-N2a appears as a very sensitive, specific functional assay [3,12]. However, the numerous protocols found in the literature highlight the lack of a consensus standardized protocol for CBA-N2a. It should be noted that the same observation applies to other analytical methods used for CTXs determination in biological samples such as LC-MS/MS and references therein, [99–101]. In this context, the present work aimed at revisiting several key parameters of the CBA-N2a as a step towards the standardized and reliable detection of two groups of toxins frequently involved in toxic outbreaks, namely VGSC activators (e.g., CTXs, PbTxs) and inhibitors (e.g., STX, dc-STX).

3.1. Revisit of the CBA-N2a

Overall, six key parameters of the CBA-N2a were revisited.

The first parameter was the cell layer density reached in microplate wells following a 22 to 26 h growth period at 37 °C, as assessed by measuring the absorbance values of the cell layer. The growth curve of the N2a cell line used in this study was established, demonstrating that the maximum number of cells that can be supported in 0.32 cm^2 wells could not exceed ≈ 100,000 cells. Our data also showed that seeding microplates at 50,000 cells/well allowed reaching this high cell density after ≈ 22 h of culture, with 100% confluence of cell layer in wells, consistent with results obtained in previous studies [59,77]. Obtaining maximum confluence with cells in late log growth phase ensures optimal absorbance data [23]. In the literature, however, a cell layer confluence > 90% was reportedly reached using different cell seeding densities, e.g., 10,000 cells/well [63], 30,000 cells/well [64,73] and 100,000 cells/well [19,20,47,48]. Some authors reported higher cell seeding densities of 250,000 cells/well [57,71], but such assay conditions do not appear physically or physiologically relevant. Actually, our results suggest that it is not the cell seeding density in itself that matters when seeking to achieve an optimal implementation of cell layer in culture microplates, and that this parameter should be adjusted to both the growth rate of the N2a cell line and the culture conditions used in the different laboratories. Therefore, growth experiments should be conducted in each laboratory in order to characterize the growth pattern of the cell line(s) in use.

The second parameter examined in this study was the MTT incubation time for cell viability assessment [19]. Here, the MTT assay was standardized using an optimal wavelength of 570 nm and an incubation time of 45 min, allowing to obtain reproducible absorbance values ≥ 1 when a maximum density of adherent cells is reached in wells after 26 h of growth, i.e., ≈ 100,000 cells/well. In all further experiments, this condition which serves as an indicator of the good health conditions of N2a cells before exposure to toxic extracts was referred to as the "initial cell viability", and was assessed by implementing a RCV control microplate. Moreover, setting the initial cell viability to an absorbance value ≥ 1 (=baseline viability) allowed us to establish a full sigmoidal dose-response curve derived from testing an eight points serial dilution of the toxin standard/sample. Any subsequent drop in cell viability following O/V treatment and toxin exposure, could then be most likely attributed to the specific effect of toxin standard/biological sample, and not from non optimal culture conditions or oversensitivity of cells.

Regarding the selection of a wavelength at 570 nm, our results are in agreement with previous work showing that the peak absorbance of MTT formazan precipitate is classically observed at 562 nm with shoulders at 512 and 587 nm [29]. Wavelengths within this range have been used in a number of studies, [36,42,50,64,67,87,90], and were generally associated with MTT incubation time around 30 min. However, wavelengths outside this range were also tested, resulting in a weaker optical signal [29] consistent with the fact that the use of lower [45,76,84,85] or higher wavelengths [57,61,79,93] often required longer MTT incubation times (>60 min).

The third parameter analyzed was the final viability of N2a cells under OV^- conditions, following an additional culture period of 19 h, overnight. Indeed, the N2a growth curve previously established showed that a loss in N2a cell viability is likely to occur at this stage when a high cell density is reached. To prevent this, a complete renewal of the used growth medium with a 2% FBS culture medium was found sufficient to stabilize cell viability throughout the CBA-N2a, as evidenced by absorbance data that were consistently found ≥1 in COV$^-$ control wells. This operation appears all the more relevant since MTT assay does not actually measure the number of viable cells or their growth, but the activity of an integrated set of enzyme linked to cell metabolism [24]. The necessity of this medium renewal has been previously highlighted in many studies [34,49,51,56,71,79,80,87,93] although the benefits of such a procedure were never clearly explained.

The fourth parameter revisited was the concentrations of Ouabain and Veratridine applied in O/V treatment. In this study, we were able to determine two distinct ranges of O/V concentrations to use depending on the group of toxins targeted in the CBA-N2a, i.e., VGSC activators and inhibitors. For the specific detection of VGSC activators in which a loss in cell viability is sought, non-destructive concentrations of O/V with respect to N2a cell viability ranging from 80/8 and 100/10 μM are recommended in order to maintain a final cell viability in COV$^+$ control close to the initial cell viability of RCV control (absorbance values ≥ 1). Conversely, for the specific detection of VGSC inhibitors in which a restoration of cell viability is sought, destructive concentrations of O/V ranging from 270/27 to 300/30 μM should be used in order to induce a residual cell viability in COV$^+$ control corresponding to approximately 10% (absorbance value of ≈ 0.1) of the initial cell viability measured in the RCV control. Moreover, the ranges of these two types of O/V treatments seem to induce constant effects on N2a cells although high passages were used and that these latter were distant from each other up to 420 passages. However, passage specific effects were not tested directly in this study, and may be the focus of future work. These findings are currently verified on the N2a cell line used in our laboratory at lower number of cell passages. The 500/50 μM O/V treatment originally proposed by Jellett [17] to detect STX was in fact chosen to induce "a lysis of many of the cells in 24 h incubation time". Such characteristics reinforce the idea that the CBA-N2a is a bioassay that is function-specific rather than toxin-specific [102].

The use of 83/17 μM [50] or 100/10 μM [83] in O/V treatment has been reported in the literature for the detection of PbTxs, although a 10–20% reduction in cell viability was observed. Similarly, a 100/10 μM O/V treatment is often selected for the detection of CTXs in *Gambierdiscus* spp. extracts [56,65,79,80,93] and fish samples [54,55,57,71,74,80], although in many of these studies no information on whether a renewal of the growth medium was applied prior to O/V treatment or on the efficiency of O/V treatment was provided. Consistent with our results is the report of the use of a 300/30 μM O/V treatment in previous studies aiming at detecting STX, which resulted in an 80% reduction in cell viability [83,94]. Additionally, previous studies showed a reduction of cell viability of 40% and up to 69–92% without and following the renewal of the growth medium, respectively, under O/V treatment at 500/50 μM [34,96]. In the present study, an O/V treatment at 300/30 μM was sufficient to induce a 100% reduction in cell viability, suggesting that the renewal of the medium was beneficial to cell physiology, thus allowing to optimize the effect of O/V treatment. However, the extent to which reduction of FBS in the culture medium from 5% to 2% following medium renewal might have rendered the cells more sensitive to toxins remains to be tested in follow-up experiments. The direct addition of O/V treatment in old culture medium following cell layer implementation in wells [19] should be avoided to ensure appropriate preservation of cell viability and further O/V treatment efficiency.

The fifth parameter examined was the concentration of solvents used to resuspend toxin standard solutions or dry extracts. This study showed that, for MeOH and DMSO, the highest final concentrations should be 0.5% to avoid non-specific cytotoxic effects on N2a cell viability, which is ensured by a dilution of at least 1:200 of the sample stock solutions. These possible solvent interferences were addressed in various ways in previous studies, including working at a final solvent concentration of 0.25–0.3% [11,87], performing an additional evaporation step before re-dissolution of toxins or biological samples in RPMI medium [54,74,81,91,103], or adding 1% MeOH in controls wells to identify any cytotoxic effects [64,73] or 5% MeOH known to induce no more than 20% of cytotoxicity on cell lines [53].

The sixth and last parameter was the maximum amount of extract to expose to avoid potential matrix effects (MCE) as compounds that often co-extract with the analyte(s) of interest can induce unspecific effects on N2a cell viability [81]. In this study, this value was determined at 10,000 pg of dry extract/mL based on the LF90/10 dry extract weight (DEW) instead of the fresh tissue equivalent (FW) used to prepare these extracts, given the significant differences often observed in the DEW/FW ratio of biological samples. Interestingly, only one previous study has used this rationale and has calculated the MCE based on lipid extracts [84]. It should be noted that MCE values are likely to vary according to the nature of tissue analyzed (e.g., flesh, viscera, fin, etc.), the different trends in lipid and water contents in tissue when sampled at different stages of the life history of a given fish species [104]. Moreover, MCE also depends on the extraction protocols and extraction efficiency [46,57,73,105], as shown for CTXs for which no reference consensus extraction protocol is available as yet [105]. Thus, until a standardized universal protocol is made available, researchers conducting similar studies should necessarily perform their own matrix assessments before applying CBA-N2a.

3.2. Performance of the Revisited CBA-N2a

The accuracy of the data provided by this revisited CBA-N2a, relies, in part, on the availability of five viability controls, i.e., RCV, COV^-, COV^+, $QCOV^-$ and $QCOV^+$ that were established throughout the various stages of CBA-N2a in order to verify: (i) the initial (RCV control) and final viability of N2a cells (COV^- controls), (ii) the efficiency of O/V treatment (COV^+ controls) and (iii) the detection of the specific mode of action of VGSC activators vs. inhibitors ($QCOV^-$ and $QCOV^+$ controls). If the use of percentages is recurrent in most studies to express cell viability results [7,8,32,33,36,37,43,46,53,59,62,63,66–68,70,73,74,76,81,83,85,87,88,90,91,95,97,103,106–112], the originality of our revisited CBA-N2a is to recommend the use of absorbance data by establishing a RCV control that characterizes the baseline viability of N2a cell layer prior to O/V treatment and toxin exposure. This RCV control serves as a reference from which all other viability controls as well as viability data of standard and samples depend. This RCV control could be considered as a common reference across all laboratories employing this method.

As a result of this comprehensive revisit of the CBA-N2a, a practical guide with a step-by-step protocol was defined (Supplementary Materials), that is accessible to both supervised beginners and experienced users. It also provides useful decision limits when interpreting CBA-N2a data in the framework of routine CBA-N2a based monitoring programs.

In this study, the sensitivity of the CBA-N2a was characterized by EC_{50} values found for P-CTX3C at 1.7 ± 0.35 pg/mL, consistent with previously published values of 1.3 ± 0.06 pg/mL [46], 1.66 ± 0.16 pg/mL [64,73] and 1.44 ± 0.70 pg/mL [77,78]. However, quite different values can also be found in the literature, e.g., 0.57 ± 0.11 pg/mL [90], 0.914 ± 0.127 pg/mL [36] and 3.10 ± 0.76 pg/mL [59]. These discrepancies can be attributed to different experimental conditions including different concentrations of O/V treatment [73] or the use of MeOH as a sample vehicle for higher sensitivity [90]. The EC_{50} found for PbTx3 was 5.8 ± 0.9 ng/mL, which is lower than the previously published value of 65.60 ± 23 ng/mL [37].

When comparing the mean EC_{50} values of P-CTX3C to those obtained via other functional assays, they were found 350-fold lower than those estimated via the radioactive receptor binding assay (rRBA), i.e., 0.62± 0.16 ng/mL [113] and 0.61 ± 0.01 ng/mL [46], and the fluorescent RBA, i.e., 0.66 ± 0.16 ng/mL [73], respectively. For PbTx3, EC_{50} values were 2-fold higher than those derived from rRBA experiments (2.77 ± 1.09 ng/mL [37] and 2.06 ± 0.16 ng/mL [113]). These discrepancies observed between functional assays can be explained by the fact that the activity of these polyether toxins depends not only on their mode of action, and their affinity for specific sodium channel isoforms [114], but also on their efficacy on other ion channels and references therein, [37,115–117]. As for the detection of STX and dc-STX, the comparison of CBA-N2a results obtained in this study with other functional assays was not possible since previous studies used different units.

For VGSC inhibitors, STX and dc-STX displayed typical dose-response curves with positive Hillslope in the presence of lethal 270/27 μM OV treatment giving EC_{50} values of 3 ± 0.5 and

15.8 ± 3 ng/mL for STX and dc-STX, respectively. At the EC_{20}, this revisited CBA-N2a could specifically detect inhibition of VGSCs at 1 and 5 ng/mL for STX and dc-STX, respectively. This result is consistent with a detection limit established at 2 ng/mL for STXs by Manger et al. [19].

This improved CBA-N2a was further applied to the detection of CTX-like toxicity in fish matrix giving mean LOD and LOQ values of 0.031 ± 0.008 and 0.064 ± 0.016 ng P-CTX3C eq/g fish flesh respectively. In order to compare these results with those obtained via analytical techniques such as LC-MS/MS data, this LOD value was further expressed in P-CTX1B equivalent, taking into account a toxicity equivalent factor (TEF) of 0.2 for P-CTX3C [118]. A LOD value of 0.0062 ng P-CTX1B eq/g was found below the recommended threshold of 0.01 ppb for P-CTX-1B [118,119], indicating that both CBA-N2a and LC-MS/MS share similar level of sensitivity for the detection of P-CTXs [61,100,120–122]. However, chromatographic analyses remain unavoidable for the formal characterization of the different structural analogs (toxin congeners) present in biological samples.

Based on EC_{50} data, the toxin content in four fish samples of known ciguatoxic status was further checked via the CBA-N2a. Only two fish specimens, namely *Chlorurus microrhinos* (Cmic19) and *Epinephelus merra* (Emer05) were found to contain 6.66 ± 0.68 and 3.31 ± 0.35 ng P-CTX-3C eq/g fish flesh, respectively, corresponding to ≈ 1.34 and 0.65 ng P-CTX1B eq/g, which are 134- and 65- fold above the advisory level recommended by both the US FDA and EFSA [118,119]. The composite cytotoxicity detected in the Cmic19 *Chlorurus microrhinos* is in agreement with fRBA and LC-MS/MS analyses, which confirmed the presence of six distinct P-CTX congeners in this fish [120]. Interestingly, these toxicity data also confirmed that the overall toxicity of herbivorous fish (parrot-fish) can sometimes surpass that of a carnivorous fish (grouper), although CTX toxin profiles occurring in individuals with different trophic habits can differ significantly owing to the biotransformation processes that occur along the food chain [120,123,124]. In addition to biotransformation, several other traits can be responsible for the unique CTX toxin profiles of individual fish, including site-specificity, feeding behavior and ontogenetic dietary shifts [61,123,125].

Several intra- and/or inter-assay comparisons of the revisited CBA-N2a were performed to assess the coefficients of variations (CVs) obtained for key parameters of CBA-N2a dose-response curves, and evaluate the repeatability and reproducibility of the method. Results indicate that for both VGSC activators and inhibitors, EC_{50} values showed CVs ranging from 10% to 26%, which are considered as acceptable values for functional tests. These results are congruent with previous studies showing 3.5 to 25% and 5 to 24.7% of CBA-N2a variability for intra and inter-assay, respectively [46,80,93,126]. A variability of up to 30% is generally admitted for other functional tests [127–129]. Additionally, no significant differences were found for EC_{50} and EC_{80} values of P-CTX3C and PbTx3 standards obtained from two non destructive O/V treatments run at distinct cell passages. Using different O/V concentrations selected in the range of 80/8–100/10 μM and different numbers of cell passage seem to not impact the values of EC_{50} and EC_{80} in our study. For LOD and LOQ values, CVs below 29% were obtained vs. CV of 13% for composite toxicity estimates in the two ciguatoxic fishes.

3.3. *Advantages of the Modified Protocol*

An advantage of the CBA-N2a is necessary material and reagents, as well as basic laboratory equipments are readily (commercially) available.

Several opportunities to reduce reagents were identified at different steps of the test, e.g., using 5% instead of 10% FBS culture medium for the seeding of N2a cells, renewing culture medium with only 2% FBS growth medium as opposed to 5% or 10% FBS commonly used in many studies [34,49,51,56,71,79,80,87,93], or working at lower O/V concentrations for O/V treatment than those originally proposed [83]. Moreover, the direct addition of O/V solution into the renewed growth medium also contributes to a better homogeneity of O/V deposit for increased repeatability and reproducibility, and a reduction of variability sources by limiting the number of pipetting. Finally, this revisited protocol allowed the use of reduced amounts of expensive toxin standards with e.g., only 80 pg of P-CTX3C toxin standard required per microplate to establish a full dose-response curve.

Another advantage of this improved test is its high modularity with respect to the specific detection of a wide range of toxins acting on VGSCs, i.e., both VGSC activators and inhibitors, thanks to the definition of O/V treatment conditions allowing the visualization of a drop in cell viability vs. cell layer restoration, respectively. This protocol also offers the possibility to assay as many as 15 microplates in one experiment (in addition to the one used for RCV control) that can be used either for the qualitative screening of 120 distinct samples at a single concentration (Figures S3 and S4), or the composite toxicity analyses of 14 samples and one toxin standard tested at eight distinct concentrations (Figures S5 and S6), depending on the laboratory research goals. For instance, the high throughput screening approach would be more suitable in risk monitoring programs aiming at the random testing of a high number of wild fish specimens, e.g., to confirm the bioaccumulation of CTXs in the marine fauna and/or the emergence of CP in novel areas. The composite toxicity estimation would be preferred when confirmatory analyses in fish products are required, e.g., to confirm the diagnosis of CP in patients, or help link the CTX content in fish to the symptoms and/or severity of the poisoning. Alternatively, using a two-tiered approach in monitoring programs is also feasible, i.e., perform rapid screening tests on a whole batch of fish specimens, and then conduct composite toxicity estimations only on those specimens found suspect or positive.

3.4. Limitations and Gaps of the Present Study

Despite the significant improvements described in the present study, the effects of several parameters that were not tested here remain to be evaluated prior to the standardization and widespread use of this method by different laboratories. For example, parameters inherent to slight differences in N2a cell lines across laboratories due e.g., to mycoplasma infection of cell cultures [130,131], cell passage numbers [132,133], desensitization treatment [134] or the use of different mammalian cell lines [53,66,83,135] should be considered in follow-up studies, as they are likely to affect cell growth rates and cell responses to toxins in this functional assay. Additional studies about the effects of reagents, media exchange, media renewal, etc. are also needed. Likewise, the inclusion of a limited number of fish in the matrix interference study presently limits the application of the method for general detection of VGSC toxins in other matrices (e.g., screening of PbTxs and STXs in shellfish samples). Another major issue is the current lack of a duly validated standardized extraction protocol for the detection of CTXs [100,105]. To this respect, the MCE value determined here must not be regarded as a universal dose as MCE closely depends not only on the extraction protocol used but also on the nature of the extract being tested [105]. Therefore, laboratories would need to perform their own matrix assessments prior to CBA-N2a studies.

4. Conclusions

Today, CBA-N2a is widely used for the detection of CTXs in a variety of biological samples, other than fish, including *Gambierdiscus* cells [46,59,64,65,79,91,93,136], giant clams [59,137], gastropods [77], sea urchins [46,78], lobsters and crabs [61], sharks [103] and even samples derived from passive monitoring sampling devices [138–140], highlighting the benefit of incorporating this functional assay into routine ciguatera risk monitoring programs for increased food security of populations worldwide. However, due to the lack of a validated reference detection method, this toxin group is still unregulated. Owing to the good performance of this revisited CBA-N2a, this improved method could reasonably be regarded as a first step towards the implementation of a reference functional detection test, provided its further validation through inter-laboratory studies. To this end, significant progress needs to be achieved in addressing several other issues such as the current shortage of certified standards and reference materials as well as the lack of consensus extraction protocols.

5. Materials and Methods

5.1. Biological Material

5.1.1. Neuroblastoma Cell Line Culture

Mouse neuroblastoma (N2a) cell line (CCL-131) was purchased from the American Type Culture Collection (ATCC, Manassas, VA, USA) (Table S4). The complete growth medium consisted of RPMI medium 1640 with HEPES and without L-Glutamine, supplemented with 10% fetal bovine serum (FBS), 2 mM GlutaMAX, 1 mM sodium pyruvate, 50 µg/mL streptomycin with 50 units/mL penicillin and 2.5 µg/mL Amphotericin B. This cell line was routinely maintained at 37 °C in a 5% CO_2 humidified atmosphere by splitting cell culture by 1:10 (every two days) or 1:30 (every three days) by rinsing the cell layer with Dulbecco's Phosphate Buffered Saline without $CaCl_2$ and $MgCl_2$ (DPBS-1X), followed by a dissociation step using 0.05% Trypsin-EDTA. The N2a cell line used in this study showed a stabilized growth after 300 P. All experiments presented in this study were done using cell passages ranging from 383 to 810 P.

All reagents are listed in Supplementary Materials (Table S5). N2a cell line was sub-cultured in 25 cm^2 or 75 cm^2 tissue culture flasks (Nunc™ Easy-Flask, ThermoFisherScientific, Kamstrupvej, Danemark), while 96-well (0.32 cm^2/well) flat bottom microplates (Falcon®, Corning Brand, New York, NY, USA) were used for CBA-N2a experiments. The list of equipments and materials are presented in Supplementary Materials (Table S6).

5.1.2. Reagents and Toxin Standards

Ouabain octahydrate (O3125), Veratridine (V5754), MTT, DMSO and MeOH are listed in Supplementary Materials (Table S4). Aqueous stock solutions of Ouabain and Veratridine were prepared at 20 mM in pure water and 5 mM in pH2 pure water, respectively.

Four toxin standards were used to characterize the mode of action of two toxin families using the CBA-N2a (Table S4): (i) two VGSC activators for which certified reference material is not available as yet, i.e., Pacific ciguatoxin P-CTX3C obtained from the bank of standards of Institut Louis Malardé [77,78,141] and brevetoxin PbTx3 (ref. L8902) purchased from Latoxan (Valence, France). Stock solutions of P-CTX3C and PbTx3 were prepared in DMSO at 20 ng/mL and 100 µg/mL, respectively. Non volatile DMSO solvent was chosen to ensure stable toxin concentrations over long periods of storage; (ii) two VGSC inhibitors for which certified reference materials were purchased from the National Research Council Canada (NRCC, Halifax, NS, Canada), i.e., saxitoxin STX (ref. CRM-STX-f) supplied at a concentration of 66.3 µM in aqueous hydrochloric acid 3 mM and decarbamoylsaxitoxin dc-STX (ref. CRM-dc-STX-b) supplied at a concentration of 65 µM in aqueous hydrochloric acid 3 mM. Concentrations of the STX and dc-STX standard solutions correspond to 24.7 and 21.4 µg/mL, respectively.

5.1.3. Fish Samples

The four fish samples tested in this study were collected from two ciguatera-endemic areas of French Polynesia, namely Tikehau Island (Tuamotu Archipelago) and Mangareva Island (Gambier Archipelago), and conditioned in the form of fillets kept at −20 °C until further CTX extraction. For some of them, their toxicity was also previously characterized by means of the fluorescent Receptor Binding Assay (fRBA) and/or Liquid Chromatography coupled with Tandem Mass Spectrometry (LC-MS/MS) [73,120]. Table 8 details their geographic origin, species, trophic status, as well as ciguatoxic status.

Table 8. List of fish samples tested for the revisited CBA-N2a.

Genus	Species	Feeding Type	Site	Sample	fRBA **	LC-MS/MS
Chlorurus	*microrhinos*	Herbivorous	Tikehau	Cmic02	NT *	NT *
			Mangareva	Cmic19	7.04 [73]	six P-CTXs [120]
Epinephelus	*merra*	Carnivorous	Mangareva	Emer05	4.4 ± 1.1 ***	NT *
			Mangareva	Emer13	NT *	NT *

* NT: not tested; ** Fluorescent RBA expressed in ng P-CTX-3C eq/g fish flesh; *** Unpublished data.

5.2. Extraction Procedure

Fish fillets were carefully homogenized in a blender (Groupe SEB, Lourdes, France) waste disposal unit for 1–2 min [113]. For each fish sample, a portion of 10 g was processed following the protocol described in Darius et al. 2018 [77]. Briefly, a liquid/liquid partition was applied followed by a purification step using Sep-Pak C18 cartridges (360 mg sorbent per cartridge; Waters®, Saint-Quentin, France) leading to three distinct liposoluble fractions, i.e., LF70/30, LF90/10 and LF100. Fractions LF90/10 likely to contain CTXs were further dried in a SpeedVac concentrator (ThermoFisherScientific, Waltham, MA, USA) and weighed with a Sartorius Micro balance (model MC 410 S, Sartorius, Göttingen, Germany) with a reading accuracy of 0.1 mg (Table 9). The resulting dry extracts were resuspended in pure methanol at a concentration of 10 mg/mL and kept at −20 °C until further CBA-N2a analysis. Prior to dosing by N2a cells, fish extracts were brought to room temperature.

Table 9. Comparison of LF90/10 dry extract weights prepared from 10 g (fresh weight) of fish samples tested with CBA-N2a.

Sample ID	Size (cm)	Weight (g)	Fresh Weight (FW) (g)	Dry Extract Weight (DEW) (mg)
Cmic02	37	926	10	3.6
Cmic19	45	2230	10	4.3
Emer05	22	185	10	3.1
Emer13	18	80	10	2.7

5.3. Neuroblastoma Cell-Based Assay (CBA-N2a)

5.3.1. Characterization of N2a Cell Growth and Initial Viability

The initial viability of N2a cells is a function of the cell density reached in wells after 26 h growth, their physiological state and the time allocated for the metabolization of MTT by viable cells.

First, in order to determine the maximum cell density supported by each of the wells (representing a surface of 0.32 cm^2) of 96-well Falcon® microplates, the growth curve of the N2a cell line used in this study was examined over a period of 4 days. For more convenience, growth experiment was conducted in 25 cm^2 culture flasks. Twelve culture flasks were seeded with 335,000 cells resuspended in 11 mL 10% FBS culture medium solution, and left to incubate for 22 h as described in Section 5.1.1. Cell densities were then monitored by sacrificing one flask every 6 h starting from 22 to 76 h. The two last flasks were sacrificed at 85 and 100 h, respectively. Cell enumeration was achieved by manual counts (n = 10/flask) using KOVA Glasstic slides (Hycor), and the results plotted against time. Cell density data were normalized to a surface of 0.32 cm^2 and used as a proxy of cell densities in 96-well microplates: e g., using this rationale, the initial cell density in a 96-well microplate was estimated at 4288 cells per well.

Second, the initial viability of N2a cells was examined under two different % FBS (5% and 10% in culture medium) and six different MTT incubation times (15–65 min). Cell numbers before seeding were determined from n = 5 counts performed on an aliquot of the cell stock suspension and apply for all further experiments. To this end, twelve microplates (six microplates per % FBS) were seeded with 200 μL of ten different cell seeding densities ranging from 10,000 to 100,000 cells/well (n = 6 wells per cell density). After 26 h growth, cell viability was assessed via the MTT assay conducted by sacrificing one microplate at 15 min and the remaining every 10 min. MTT and DMSO conditions were used as originally described [19]. The MTT protocol followed the one described in Darius et al. [77]. Briefly, after removing the culture medium from the microplate, the 60 inner wells were filled with 60 μL of MTT solution at 0.83 mg/mL prepared in 2% FBS culture medium. After MTT incubation at 37 °C in a 5% CO_2 incubator, the wells were emptied and the 60 inner wells and 12 outer wells (rows 1 and 12) were filled with 100 μL of DMSO. Following cell layer lysis with DMSO and manual homogenization, absorbance data were measured at 570 nm using an iMark™ microplate reader (Biorad, Marnes la Coquette, France). All viability data were expressed in absorbance data.

5.3.2. Characterization of N2a Cell Final Viability in the Absence and under O/V Treatment

Ouabain and Veratridine (O/V) treatment is required for the successful detection by CBA-N2a of toxins active on VGSCs. The N2a cell final viability, i.e., in the absence of O/V treatment (OV^- conditions) or following O/V treatment (OV^+ conditions) was characterized as follows.

To characterize the final viability of N2a cells in OV^- conditions, six microplates were seeded with 200 μL of a 5% FBS culture medium at ten different cell densities ranging from 10,000 to 100,000 cells/well (n = 6 wells per cell density) and left to grow for 26 h. One microplate, defined as the RCV control, was used to check the initial viability of N2a cells after 26 h growth and measured by MTT assay, while the five remaining microplates were treated as follows: culture medium was discarded under sterile conditions by overturning, and wells were dried by tapping each microplate on an ethanol-sterilized absorbent paper to remove any residual liquid. The 60 inner wells of the five microplates then received 200 μL/well of growth medium supplemented with 1%, 2%, 3%, 4% or 5% FBS, respectively, while the peripheral wells received the same volume of sterile distilled water. Culture microplates were further incubated overnight (Section 5.1.1) for an additional 19 h. Cell final viability was assessed using the MTT assay after 45 min incubation with MTT.

To characterize the final viability of N2a cells in OV^+ conditions, two microplates were seeded with 200 μL/well of a 5% FBS culture medium at an initial cell density of 50,000 ± 10,000 cells/well, and left to grow for 26 h. After cell layer settlement, one microplate served as RCV control and measured by MTT assay. For the second one, 20 mM (O) and 5 mM (V) solutions were first diluted in 2% FBS culture medium to obtain a 360/36 μM O/V stock solution, which was further used to prepare serial dilutions ranging from 340/34 to 20/2 μM (with a decrease of 20/2 μM between each dilution). After removal of the used culture medium as previously described, each of the 60 inner wells received 200 μL/well of culture medium at nineteen distinct O/V concentrations ranging from 0/0 to 360/36 μM (n = 3 wells per O/V treatment conditions, except for 0/0 μM condition for which n = 6). Peripheral wells received the same volume of sterile distilled water. Culture plates were then incubated overnight for 19 h. Cell final viability was assessed using the MTT assay after 45 min incubation with MTT. Five independent experiments were performed.

5.3.3. Characterization of the Unspecific Effects of Solvent and Dry Extract on N2a Cell Viability

The effects on N2a cell viability of two solvents commonly used to resuspend toxin standards and dry extracts, i.e., MeOH and DMSO, were examined in the absence vs. under non-destructive O/V treatment conditions. In this experiment, the 60 inner wells of three 96-well microplates were seeded with 200 μL/well of a 5% FBS culture medium at an initial cell density of 50,000 ± 10,000 cells/well, and left to grow for 26 h. One microplate served as RCV control and was measured by MTT assay while the two remaining ones were treated as follows: after the complete removal of the used culture medium as previously described, the 30 inner wells on the upper half of the microplate received 200 μL/well of a 2% FBS culture medium (OV^- conditions), while the 30 inner wells on the bottom half of the microplate received 200 μL/well of a 105/10.5 μM O/V treatment (OV^+ conditions). In parallel, 100 μL of eight-points serial dilution at 1:2 of each solvent were prepared in the same culture medium using a U-bottom 96-well microtiter plate. Then, 10 μL of each solvent dilution were directly added in triplicate wells and tested under OV$^-$ conditions versus OV$^+$ conditions (100/10 µM final concentrations). Hence, the final concentrations tested ranged from 0.037 to 4.762%. Addition of 10 µL of 2% FBS culture medium in triplicate wells under OV$^-$ and OV$^+$ conditions (COV$^-$ and COV$^+$ controls, respectively), allowed to check final viability in solvent-free growth medium. Peripheral wells received the same volume of sterile distilled water. Culture plates were left to incubate overnight for 19 h, and the cell final viability determined using the MTT assay after 45 min incubation with MTT.

In order to determine the MCE of the four fish samples, increasing concentrations of fish extracts were tested in OV$^-$ conditions only, as no activity on VGSCs is expected to occur in this condition of treatment [81]. In most CBA-N2a studies however, these concentrations are established based on the fresh tissue weight of biological samples. Here, the tested concentrations were estimated based on the DEW instead, since dry extracts interact directly on cell layers and substantial differences in DEWs are often observed between extracts prepared from a same amount of tissue sample. In order to determine the MCE of the four fish samples used in this study, the 60 inner wells of three 96-well microplates were seeded with 200 µL of a 5% FBS culture medium at an initial cell density of 50,000 ± 10,000 cells/well, then left to grow for 26 h. One microplate served as RCV control and was measured by MTT assay, while the two remaining ones were treated as follows: first, 5% FBS growth medium was renewed by addition of 200 µL of a 2% FBS culture medium in wells. In parallel, a 1:10 dilution of the LF90/10 dry extract of four fish samples was prepared in 2% FBS culture medium using a U-bottom 96-well microtiter plate, followed by a nine-points serial 1:2 dilution (v = 100 µL per concentration). Then, 10 µL of each concentration was directly added in triplicate wells, leading to final dry extract concentrations that ranged from 186 to 47,619 pg/µL, i.e., 0.517 to 132.3 µg fish flesh equivalent/µL for Cmic02, 0.433 to 110.7 µg/µL for Cmic19, 0.600 to 153.6 µg/µL for Emer05 and 0.689 to 176.3 µg/µL for Emer13. Two fish samples were tested per plate. Controls wells (COV$^-$) were also established by addition of 10 µL of 2% FBS culture medium in the absence of dry extract. Peripheral wells received the same volume of sterile distilled water. Culture plates were left to incubate overnight for 19 h, and the cell final viability determined using the MTT assay after 45 min incubation with MTT. Three independent experiments were performed.

5.3.4. Detection of VGSCs Activators and Inhibitors by CBA-N2a

In this experiment, 60 inner wells of five 96-well microplates were seeded with 200 µL of a 5% FBS culture medium at an initial cell density of 50,000 ± 10,000 cells/well, and left to grow for 26 h. One microplate served as RCV control and was measured by MTT assay, while the four remaining ones were treated as follows: first, the growth medium was renewed by the addition of 200 µL of 2% FBS culture medium in OV$^-$ conditions (upper half of the microplate) versus 200 µL of culture medium in OV$^+$ conditions (bottom half). The initial O/V concentrations in wells were 105/10.5 and 284/28.4 µM when detecting VGSC activators and VGSC inhibitors, respectively. Further, a nine-points serial 1:2 dilution of each toxin standard stock solution was prepared in 2% FBS culture medium (v = 100 µL per concentration) using a U-bottom 96-well microtiter plate, then 10 µL of each toxin concentration were directly added in triplicate under OV$^-$ and OV$^+$ conditions (Section 5.3.3). The final concentrations of toxins tested ranged from 0.074 to 19.048 fg/µL for P-CTX3C, 372 to 95,238 fg/µL for PbTx3, 368 to 94,095 fg/µL for STX and 1592 to 407,619 fg/µL for dc-STX. The final O/V concentrations in wells were 100/10 and 270/27 µM when detecting VGSC activators and VGSC inhibitors, respectively. Appropriate controls in both conditions of O/V treatment, COV$^-$ and COV$^+$, were established to verify the cell layer viability and the effect of O/V treatment, respectively, in the absence of toxins (Figure S7). The resulting full dose-response curves were used to characterize the response typical of a given standard toxin and for further toxin quantification. For each toxin standard, one microplate in three independent experiments was examined for the purpose of inter-assay variability comparison.

5.3.5. Detection of VGSC Activators in Fish Samples by CBA-N2a

This experiment was conducted as previously described in Section 5.3.4, except that the initial concentrations of O/V in wells was set to 90/9 μM. Toxin detection and quantification in biological matrix of unknown varying toxicity require the implementation of additional quality check controls (QC) in OV^- and OV^+ conditions, namely $QCOV^-$ and $QCOV^+$, to check for the validity of further toxicity results (Figure S2). Practically, in these controls, a known concentration of a VGSC activator is tested whose effect on N2a cell viability has been pre-established. The $QCOV^-$ and $QCOV^+$ were established by adding 10 μL of 0.1 μg /mL of PbTx3 in triplicate, to reach a final concentration of 4760 fg/μL of PbTx3 in wells (PbTx3 was preferred to P-CTX3C as it is commercially available). An eight-points serial 1:2 dilution of P-CTX3C and fish dry extracts were prepared (v = 100 μL per concentration using a U-bottom 96-well microtiter, then 10 μL of each concentration were directly added in triplicate under OV^- and OV^+ conditions (85.7/8.57 μM final concentrations). Hence, the final concentrations of P-CTX3C tested ranged from 0.099 to 12.70 fg/μL and from 74.4 to 9523.8 pg of dry extracts/μL for Cmic02 and Emer13, 14.9 to 1904.8 pg of dry extracts/μL for Cmic19 and 18.6 to 2381 pg of dry extracts/μL for Emer05. The full sigmoidal dose–response curves obtained for each toxin standard and fish samples when tested in parallel in the same experiment were used to determine EC_{50} values and infer toxin content in fish samples (Section 5.3.6). Three microplates in one experiment ($n = 3$) and one microplate in three independent experiments ($n = 3$) were examined for the purpose of intra- and inter-assay variability comparison, respectively.

Validation of the CBA-N2a results by means of appropriate viability controls are presented in Supplementary Materials (Table S4).

5.3.6. Absorbance Data and Toxin Analysis

First, for each experimental plate, all raw absorbance data were corrected by deducting the corresponding mean DMSO control absorbance data ($n = 12$) to obtain net absorbance data. Dose-response curves were then established by plotting net absorbance values vs. pure toxins or dry extract concentrations tested, using GraphPad Prism software version 8.1.2 (GraphPad, San Diego, CA, USA) based on a four parameter logistic regression model (4PL) according to the following equation:

$$Y = \text{Bottom} + (\text{Top} - \text{Bottom}/(1 + 10\hat{}((\text{Log}(EC_{50}) - \text{Log}(X)) * \text{Hillslope})) \quad (2)$$

In which Y is the net absorbance data and X is the concentrations tested (fg/μL), and EC_{50} (fg/μL) represents the effective concentration of dry extract inducing a viability half way (50%) between the basal (Bottom) and the maximal (Top) values of the curve (EC_{50}). This parameter is used to establish the toxic potency of each toxin standard.

The EC_{80} and EC_{20} values of toxin standards were inferred from dose-response curves and correspond to the effective concentration of dry extract inducing a viability 80% and 20% between the basal (Bottom) and the maximal (Top) values of the curve, respectively.

The limit of detection (LOD) and quantification (LOQ) of the CTX-like toxicity in fish samples were determined according to the following equations:

$$LOD = (EC_{80}/MCE) \quad (3)$$

$$LOQ = (EC_{50}/MCE) \quad (4)$$

where EC_{80} and EC_{50} are the values obtained for P-CTX3C toxin standard, with values expressed in ng P-CTX3C eq/mg of dry extract.

For more convenience, LOD and LOQ values can be expressed in the same unit as the one used in the advisory level recommended by the EFSA and US FDA. Calculations are based on the fresh weight of flesh tissue extracted (FW) and the corresponding dry extract weight (DEW) (Table 9), and use the following equations:

$$LOD = (EC_{80}/MCE) \times (DEW/FW) \quad (5)$$

$$LOQ = (EC_{50}/MCE) \times (DEW/FW) \quad (6)$$

In which LOD and LOQ of CTXs in biological matrix are expressed in ng P-CTX3C eq/g fish flesh.

In the same way, quantification of the composite toxicity in fish dry extracts (T), expressed in ng P-CTX3C eq/mg, was determined by comparing the EC_{50} values of P-CTX3C and fish dry extracts determined in the same experiment, using the following equation:

$$T = EC_{50} \text{ of P-CTX3C}/EC_{50} \text{ of dry extract} \quad (7)$$

The composite toxicity in biological samples (Q), expressed in ng P-CTX3C eq/g of fish flesh, is determined using the following equation:

$$Q = T \times (DEW/FW) \quad (8)$$

5.3.7. Data Analyses

Variabilities of CBA-N2a data were examined using the mean ± SD of three microplates tested in one experiment on the same day (intra-assay comparisons) or the mean ± SD of one microplate tested in three independent experiments (inter-assay comparisons). Statistical analyses were performed by means of the Wilcoxon test with significant differences considered at p-values < 0.05, using RStudio software version 1.0.153 Version 1.0.153–© 2009-2017 (RStudio, Inc., Boston, MA, USA).

First, the intra- and inter-assay variabilities of the five viability controls (Table S7), RCV, COV^-, COV^+, $QCOV^-$ and $QCOV^+$ controls were assessed from inter-assay experiments conducted at 100/10 µM run at 662-663-666 cell passages and intra- and inter-assay experiments conducted at 85.7/8.57 µM run at 810 and 795-797-798 cell passages, respectively of CBA-N2a experiments presented in Sections 2.4.2 and 2.4.3 and Table S1.

Second, the Wilcoxon test was applied to search for variability between EC_{80} and EC_{50} of P-CTX3C and PbTx3 obtained from inter-assay experiments conducted at 100/10 µM and run at 662-663-666 cell passages, and intra- and inter-assay experiments conducted at 85.7/8.57 µM and run at 810 and 795-797-798 cell passages, respectively (Table 4, Table S2).

For intra- and inter assay comparison of LOD and LOQ values of the CTX-like toxicity in fish dry extracts, three different P-CTX3C EC_{80} or EC_{50} values were compared to a unique MCE value, i.e., 10,000 pg/µL applying for all fish samples. Then, three LOD and LOQ values were obtained giving a mean ± SD with $n = 3$ for intra- and inter-assay for all fish samples (Table 5).

For intra- and inter assay comparison of LOD and LOQ values of the CTX-like toxicity in fish flesh, one specific MCE value (converted from the DEW/FW ratio) was obtained for each fish sample. Then, three different P-CTX3C EC_{80} or EC_{50} values were compared to a unique MCE value per fish sample. In the same way, three LOD and LOQ values were obtained for each fish giving a mean ± SD with $n = 3$ for intra- and inter-assays (Table 5). Additionally, the Wilcoxon test was successfully applied to search for possible variability between LOD and LOQ values among the four fish samples (Table S3).

For intra-assay comparison of toxin content in fish sample, three different P-CTX3C EC_{50} values obtained from $n = 3$ microplates were compared to three different fish EC_{50} values obtained from $n = 3$ microplates of a given fish sample run the same day. Then, cross calculation (three different P-CTX3C EC_{50} values x three different fish EC_{50} values) were done giving nine toxin content values and a mean ± SD with $n = 9$ per fish sample (Table 7).

For inter-assay comparison of toxin content in fish sample, one P-CTX3C EC_{50} value was compared to one fish EC_{50} value for each fish sample per independent experiment run at different times. Then, three toxin content values were obtained from the three independent experiments giving a mean ± SD with $n = 3$ for each fish sample (Table 7).

Supplementary Materials: The following are available online at http://www.mdpi.com/2072-6651/12/5/281/s1. Figure S1: Examples of pictures obtained from the revisited CBA-N2a, Figure S2: Suggested layouts on a 96-wells microplate at day 2 of the CBA-N2a, Figure S3: Detection (DL) and confirmation (CL) limits adopted in qualitative screening analysis of samples tested for the presence of VGSC activators (under non-destructive O/V treatment), Figure S4: Detection (DL) and confirmation (CL) limits adopted in qualitative screening analysis of samples tested for the presence of VGSC inhibitors (under destructive O/V treatment), Figure S5: Typical dose-response curve expected with a sample positive for VGSC activators, tested at eight distinct concentrations (C1–C8) adjusted to fall below the MCE, Figure S6: Typical dose-response curve expected with a sample positive for VGSC inhibitors tested at eight distinct concentrations (C1–C8) adjusted to fall below the MCE, Table S1: Comparison between assays of the absorbance data of five viability controls, Table S2: Comparison between assays of EC_{50} and EC_{80} values of P-CTX3C and PbTx3 standards under two non destructive O/V treatments (100/10 μM vs. 85.7/8.57 μM), Table S3: Comparison between assays of LOD and LOQ values (ng P-CTX3C equivalent/g fish flesh) under non-destructive O/V treatment at 85.7/8.57 μM), Table S4: List of biological materials, Table S5: List of reagents, Table S6: List of equipments and materials, Table S7: Expected results for viability controls to enable validation of the assay.

Author Contributions: Conceptualization, J.V.; Data curation, J.V. and H.T.D.; Formal analysis, J.V. and H.T.D.; Funding acquisition, M.C. and H.T.D.; Methodology, J.V. and H.T.D.; Project administration, M.C.; Resources, M.C. and H.T.D.; Supervision, M.C.; Validation, J.V., M.C. and H.T.D.; Visualization, J.V., M.C. and H.T.D.; Writing—original draft, J.V., M.C. and H.T.D.; Writing—review & editing, J.V., M.C. and H.T.D. All authors have read and agreed to the published version of the manuscript.

Funding: The present work was supported by funds from the countries of France and French Polynesia in the framework of the CARISTO-Pf" (no. 7937/MSR/REC of 4 December 2015 and Arrêté no. HC/491/DIE/BPT of 30 March 2016) and FLUOTRACK-CIGUATERA" (no. 023-15 of 19 February 2015) research programs.

Acknowledgments: The authors greatly acknowledge Philippe CRUCHET for the chemical extraction of fish samples and Sébastien LONGO for statistical analysis, as well as the three anonymous reviewers whose comments greatly helped improve the manuscript.

Conflicts of Interest: The authors declare no conflict of interest. The funders had no role in the design of the study; in the collection, analyses, or interpretation of data; in the writing of the manuscript; or in the decision to publish the results.

References

1. Daneshian, M.; Botana, L.M.; Dechraoui Bottein, M.-Y.; Buckland, G.; Campàs, M.; Dennison, N.; Dickey, R.W.; Diogène, J.; Fessard, V.; Hartung, T.; et al. A roadmap for hazard monitoring and risk assessment of marine biotoxins on the basis of chemical and biological test systems. *Altern. Anim. Exp. (ALTEX)* **2013**, *30*, 487–545.
2. Nicolas, J.; Hendriksen, P.J.M.; Gerssen, A.; Bovee, T.F.H.; Rietjens, I.M.C.M. Marine neurotoxins: State of the art, bottlenecks, and perspectives for mode of action based methods of detection in seafood. *Mol. Nutr. Food Res.* **2014**, *58*, 87–100. [CrossRef] [PubMed]
3. Nicolas, J.; Hoogenboom, R.L.A.P.; Hendriksen, P.J.M.; Bodero, M.; Bovee, T.F.H.; Rietjens, M.C.M.; Gerssen, A. Marine biotoxins and associated outbreaks following seafood consumption: Prevention and surveillance in the 21st century. *Glob. Food Secur.* **2017**, *15*, 11–21. [CrossRef]
4. Caillaud, A.; de la Iglesia, P.; Darius, H.T.; Pauillac, S.; Aligizaki, K.; Fraga, S.; Chinain, M.; Diogène, J. Update on methodologies available for ciguatoxin determination: Perspectives to confront the onset of ciguatera fish poisoning in Europe. *Mar. Drugs* **2010**, *8*, 1838–1907. [CrossRef] [PubMed]
5. Rodríguez, I.; Vieytes, M.R.; Alfonso, A. Analytical challenges for regulated marine toxins. Detection methods. *Curr. Opin. Food Sci.* **2017**, *18*, 29–36. [CrossRef]
6. European Commission. Commission regulation (EU) No 15/2011 of 10 January 2011 amending Regulation (EC) No 2074/2005 as regards recognised testing methods for detecting marine biotoxins in live bivalve mollusks. *Off. J. Eur. Union* **2011**, *50*, 3–4.
7. Nicolas, J.; Bovee, T.F.H.; Kamelia, L.; Rietjens, I.M.C.M.; Hendriksen, P.J.M. Exploration of new functional endpoints in neuro-2a cells for the detection of the marine biotoxins saxitoxin, palytoxin and tetrodotoxin. *Toxicol. In Vitro* **2015**, *30*, 341–347. [CrossRef] [PubMed]
8. Estevez, P.; Castro, D.; Pequeño-Valtierra, A.; Giraldez, J.; Gago-Martinez, A. Emerging marine biotoxins in seafood from European coasts: Incidence and analytical challenges. *Foods* **2019**, *8*, 149. [CrossRef] [PubMed]
9. Turner, A.D.; Dhanji-Rapkova, M.; Fong, S.Y.T.; Hungerford, J.; McNabb, P.S.; Boundy, M.J.; Harwood, D.T. Ultrahigh-performance hydrophilic interaction liquid chromatography with tandem mass spectrometry

method for the determination of paralytic shellfish toxins and tetrodotoxin in mussels, oysters, clams, cockles, and scallops: Collaborative study. *J. AOAC Int.* **2020**, *103*, 30. [CrossRef] [PubMed]
10. Turner, A.D.; Hatfield, R.G.; Maskrey, B.H.; Algoet, M.; Lawrence, J.F. Evaluation of the new European Union reference method for paralytic shellfish toxins in shellfish: A review of twelve years regulatory monitoring using pre-column oxidation LC-FLD. *TrAC Trends Anal. Chem.* **2019**, *113*, 124–139. [CrossRef]
11. Nicolas, J.; Hendriksen, P.J.M.; van Kleef, R.G.D.M.; de Groot, A.; Bovee, T.F.H.; Rietjens, I.M.C.M.; Westerink, R.H.S. Detection of marine neurotoxins in food safety testing using a multielectrode array. *Mol. Nutr. Food Res.* **2014**, *58*, 2369–2378. [CrossRef]
12. Fessard, V. Chapter Seven—Cytotoxicity assays: Identification of toxins and mechanism of action. In *Comprehensive Analytical Chemistry*; Diogène, J., Campàs, M., Eds.; Elsevier: Amsterdam, The Netherlands, 2017; Volume 78, pp. 231–275.
13. Kogure, K.; Tamplin, M.L.; Simidu, U.; Colwell, R.R. A tissue culture assay for tetrodotoxin, saxitoxin and related toxins. *Toxicon* **1988**, *2*, 191–197. [CrossRef]
14. Catterall, W.A.; Nirenberg, M. Sodium uptake associated with activation of action potential ionophores of cultured neuroblastoma and muscle cells. *Proc. Natl. Acad. Sci. USA* **1973**, *70*, 3759–3763. [CrossRef]
15. Kodama, M. Possible links between bacteria and toxin production in algal blooms. In *Toxic Marine Phytoplankton*; Graneli, A., Ed.; Elsevier Science Publishing Co., Inc.: Amsterdam, The Netherlands, 1990; pp. 52–61.
16. Ogata, T.; Sato, S.; Kodama, M. Paralytic shellfish toxins in bivalves which are not associated with dinoflagellates. *Toxicon* **1989**, *27*, 1241–1244. [CrossRef]
17. Jellett, J.F.; Marks, L.J.; Stewart, J.E.; Dorey, M.L.; Watson-Wright, W.; Lawrence, J.F. Paralytic shellfish poison (saxitoxin family) bioassays: Automated endpoint determination and standardization of the in vitro tissue culture bioassay, and comparison with the standard mouse bioassay. *Toxicon* **1992**, *30*, 1143–1156 [CrossRef]
18. Gallacher, S.; Birkbeck, T.H. A tissue culture assay for direct detection of sodium channel blocking toxins in bacterial culture supernates. *FEMS Microbiol. Lett.* **1992**, *92*, 101–107. [CrossRef]
19. Manger, R.L.; Leja, L.S.; Lee, S.Y.; Hungerford, J.M.; Wekell, M.M. Tetrazolium-based cell bioassay for neurotoxins active on voltage-sensitive sodium channels: Semiautomated assay for saxitoxins, brevetoxins, and ciguatoxins. *Anal. Biochem.* **1993**, *214*, 190–194. [CrossRef] [PubMed]
20. Manger, R.L.; Leja, L.S.; Lee, S.Y.; Hungerford, J.M.; Hokama, Y.; Dickey, R.W.; Granade, H.R ; Lewis, R.; Yasumoto, T.; Wekell, M.M. Detection of sodium channel toxins: Directed cytotoxicity assays of purified ciguatoxins, brevetoxins, saxitoxins, and seafood extracts. *J. AOAC Int.* **1995**, *78*, 521–527. [CrossRef] [PubMed]
21. LePage, K.T.; Dickey, R.W.; Gerwick, W.H.; Jester, E.L.; Murray, T.F. On the use of neuro-2a neuroblastoma cells versus intact neurons in primary culture for neurotoxicity studies. *Crit. Rev. Neurobiol.* **2005**, *17*, 27–50. [CrossRef]
22. Mosmann, T. Rapid colorimetric assay for cellular growth and survival: Application to proliferation and cytotoxicity assays. *J. Immunol. Methods* **1983**, *65*, 55–63. [CrossRef]
23. Carmichael, J.; DeGraff, W.G.; Gazdar, A.F.; Minna, J.D.; Mitchell, J.B. Evaluation of a tetrazolium-based semiautomated colorimetric assay: Assessment of chemosensitivity testing. *Cancer Res.* **1987**, *47*, 936–942. [PubMed]
24. Berridge, M.V.; Herst, P.M.; Tan, A.S. Tetrazolium dyes as tools in cell biology: New insights into their cellular reduction. *Biotechnol. Annu. Rev.* **2005**, *11*, 127–152. [PubMed]
25. Hansen, J.; Bross, P. A cellular viability assay to monitor drug toxicity. In *Protein Misfolding and Cellular Stress in Disease and Aging: Concepts and Protocols*; Bross, P., Gregersen, N., Eds.; Humana Press: Totowa, NJ, USA, 2010; pp. 303–311.
26. Tsukatani, T.; Higuchi, T.; Suenaga, H.; Akao, T.; Ishiyama, M.; Ezoe, T.; Matsumoto, K. Colorimetric microbial viability assay based on reduction of water-soluble tetrazolium salts for antimicrobial susceptibility testing and screening of antimicrobial substances. *Anal. Biochem.* **2009**, *393*, 117–125. [CrossRef] [PubMed]
27. Wang, H.; Cheng, H.; Wang, F.; Wei, D.; Wang, X. An improved 3-(4,5-dimethylthiazol-2-yl)-2,5-diphenyl tetrazolium bromide (MTT) reduction assay for evaluating the viability of Escherichia coli cells. *J. Microbiol. Methods* **2010**, *82*, 330–333. [CrossRef] [PubMed]

28. Abe, K.; Matsuki, N. Measurement of cellular 3-(4,5-dimethylthiazol-2-yl)-2,5-diphenyltetrazolium bromide (MTT) reduction activity and lactate dehydrogenase release using MTT. *Neurosci. Res.* **2000**, *38*, 325–329. [CrossRef]
29. Stockert, J.C.; Blázquez-Castro, A.; Cañete, M.; Horobin, R.W.; Villanueva, Á. MTT assay for cell viability: Intracellular localization of the formazan product is in lipid droplets. *Acta Histochem.* **2012**, *114*, 785–796. [CrossRef]
30. Manger, R.L.; Leja, L.S.; Lee, S.Y.; Hungerford, J.M.; Wekell, M.M. Tetrazolium-Based Cell Bioassay for Marine Neurotoxins. U.S. Patent 5,420,011, 30 May 1995.
31. Manger, R.L.; Leja, L.S.; Lee, S.Y.; Hungerford, J.M.; Wekell, M.M. Cell Bioassay for Neurotoxins. U.S. Patent 6,174,690 B1, 16 January 2001.
32. Aballay-Gonzalez, A.; Ulloa, V.; Rivera, A.; Hernández, V.; Silva, M.; Caprile, T.; Delgado-Rivera, L.; Astuya, A. Matrix effects on a cell-based assay used for the detection of paralytic shellfish toxins in bivalve shellfish samples. *Food Addit. Contam. Part A* **2016**, *33*, 869–875. [CrossRef]
33. Bienfang, P.; DeFelice, S.; Laws, E.; Wallsgrove, N.; Caldwell, P. Ciguatoxicity in the main Hawaiian Islands: Spatial and temporal variability in the introduced reef carnivore Cephalopholis Argus. *J. Res. Environ. Sci. Toxicol.* **2012**, *1*, 47–57.
34. Truman, P.; Lake, R.J. Comparison of mouse bioassay and sodium channel cytotoxicity assay for detecting paralytic shellfish poisoning toxins in shellfish extracts. *J. AOAC Int.* **1996**, *79*, 1130–1133. [CrossRef]
35. Humpage, A.R.; Ledreux, A.; Fanok, S.; Bernard, C.; Briand, J.-F.; Eaglesham, G.; Papageorgiou, J.; Nicholson, B.; Steffensen, D. Application of the neuroblastoma assay for paralytic shellfish poisons to neurotoxic freshwater cyanobacteria: Interlaboratory calibration and comparison with other methods of analysis. *Environ.Toxicol. Chem.* **2007**, *26*, 1512–1519. [CrossRef]
36. Ledreux, A.; Brand, H.; Chinain, M.; Bottein, M.-Y.D.; Ramsdell, J.S. Dynamics of ciguatoxins from Gambierdiscus polynesiensis in the benthic herbivore Mugil cephalus: Trophic transfer implications. *Harmful Algae* **2014**, *39*, 165–174. [CrossRef]
37. Bottein Dechraoui, M.Y.; Tiedeken, J.A.; Persad, R.; Wang, Z.; Granade, H.R.; Dickey, R.W.; Ramsdell, J.S. Use of two detection methods to discriminate ciguatoxins from brevetoxins: Application to great barracuda from Florida Keys. *Toxicon* **2005**, *46*, 261–270. [CrossRef] [PubMed]
38. Bottein Dechraoui, M.Y.; Wang, Z.; Turquet, J.; Chinain, M.; Darius, T.; Cruchet, P.; Radwan, F.F.Y.; Dickey, R.W.; Ramsdell, J.S. Biomonitoring of ciguatoxin exposure in mice using blood collection cards. *Toxicon* **2005**, *46*, 243–251. [CrossRef] [PubMed]
39. Bottein Dechraoui, M.-Y.; Rezvani, A.H.; Gordon, C.J.; Levin, E.D.; Ramsdell, J.S. Repeat exposure to ciguatoxin leads to enhanced and sustained thermoregulatory, pain threshold and motor activity responses in mice: Relationship to blood ciguatoxin concentrations. *Toxicology* **2008**, *246*, 55–62. [CrossRef]
40. Dechraoui Bottein, M.Y.; Kashinsky, L.; Wang, Z.; Littnan, C.; Ramsdell, J.S. Identification of ciguatoxins in Hawaiian monk seals Monachus schauinslandi from the Northwestern and Main Hawaiian Islands. *Environ. Sci. Technol.* **2011**, *45*, 5403–5409. [CrossRef]
41. Dechraoui Bottein, M.-Y.; Wang, Z.; Ramsdell, J.S. Toxicokinetics of the ciguatoxin P-CTX-1 in rats after intraperitoneal or oral administration. *Toxicology* **2011**, *284*, 1–6. [CrossRef]
42. Ledreux, A.; Ramsdell, J.S. Bioavailability and intravenous toxicokinetic parameters for Pacific ciguatoxin P-CTX-1 in rats. *Toxicon* **2013**, *64*, 81–86. [CrossRef]
43. Rhodes, L.L.; Smith, K.F.; Munday, R.; Selwood, A.I.; McNabb, P.S.; Holland, P.T.; Bottein, M.-Y. Toxic dinoflagellates (Dinophyceae) from Rarotonga, Cook Islands. *Toxicon* **2010**, *56*, 751–758. [CrossRef]
44. Loeffler, C.R.; Robertson, A.; Flores Quintana, H.A.; Silander, M.C.; Smith, T.B.; Olsen, D. Ciguatoxin prevalence in 4 commercial fish species along an oceanic exposure gradient in the US Virgin Islands. *Environ. Toxicol. Chem.* **2018**, *37*, 1852–1863. [CrossRef]
45. Kerbrat, A.-S.; Darius, H.T.; Pauillac, S.; Chinain, M.; Laurent, D. Detection of ciguatoxin-like and paralysing toxins in Trichodesmium spp. from New Caledonia lagoon: New Caledonia tropical lagoons: An overview of multidisciplinary investigations. *Mar. Pollut. Bull.* **2010**, *61*, 360–366. [CrossRef]
46. Pawlowiez, R.; Darius, H.T.; Cruchet, P.; Rossi, F.; Caillaud, A.; Laurent, D.; Chinain, M. Evaluation of seafood toxicity in the Australes archipelago (French Polynesia) using the neuroblastoma cell-based assay. *Food Addit. Contam. Part A* **2013**, *30*, 567–586. [CrossRef] [PubMed]

47. Zheng, X.; Hu, B.; Gao, S.X.; Liu, D.J.; Sun, M.J.; Jiao, B.H.; Wang, L.H. A saxitoxin-binding aptamer with higher affinity and inhibitory activity optimized by rational site-directed mutagenesis and truncation. *Toxicon* **2015**, *101*, 41–47. [CrossRef] [PubMed]
48. Dickey, R.W. Ciguatera toxins: Chemistry, toxicology and detection. In *Seafood and Freshwater Toxins. Pharmacology, Physiology and Detection*, 2nd ed.; CRC Press, Taylor and Francis Group: Boca Raton, FL, USA, 2008; pp. 479–500.
49. Hayashi, R.; Saito, H.; Okumura, M.; Kondo, F. Cell bioassay for Paralytic Shellfish Poisoning (PSP): Comparison with postcolumn derivatization liquid chromatographic analysis and application to the monitoring of PSP in shellfish. *J. Agric. Food Chem.* **2006**, *54*, 269–273. [CrossRef] [PubMed]
50. Truman, P.; Stirling, D.J.; Northcote, P.; Lake, R.J.; Seamer, C.; Hannah, D.J. Determination of brevetoxins in shellfish by the neuroblastoma assay. *J. AOAC Int.* **2002**, *85*, 1057–1063. [CrossRef]
51. Truman, P. A cellular target for the lipophilic toxins from Karenia brevisulcata. *Toxicon* **2007**, *50*, 251–255. [CrossRef]
52. Truman, P.; Keyzers, R.A.; Northcote, P.T.; Ambrose, V.; Redshaw, N.A.; Hoe Chang, F. Lipophilic toxicity from the marine dinoflagellate Karenia brevisulcata: Use of the brevetoxin neuroblastoma assay to assess toxin presence and concentration. *Toxicon* **2005**, *46*, 441–445. [CrossRef]
53. Sérandour, A.-L.; Ledreux, A.; Morin, B.; Derick, S.; Augier, E.; Lanceleur, R.; Hamlaoui, S.; Moukha, S.; Furger, C.; Biré, R.; et al. Collaborative study for the detection of toxic compounds in shellfish extracts using cell-based assays. Part I: Screening strategy and pre-validation study with lipophilic marine toxins. *Anal. Bioanal. Chem.* **2012**, *403*, 1983–1993. [CrossRef]
54. Soliño, L.; Widgy, S.; Pautonnier, A.; Turquet, J.; Loeffler, C.R.; Flores Quintana, H.A.; Diogène, J. Prevalence of ciguatoxins in lionfish (Pterois spp.) from Guadeloupe, Saint Martin, and Saint Barthélmy Islands (Caribbean). *Toxicon* **2015**, *102*, 62–68. [CrossRef]
55. Sanchez-Henao, J.A.; García-Álvarez, N.; Fernández, A.; Saavedra, P.; Silva Sergent, F.; Padilla, D.; Acosta-Hernández, B.; Martel Suárez, M.; Diogène, J.; Real, F. Predictive score and probability of CTX-like toxicity in fish samples from the official control of ciguatera in the Canary Islands. *Sci. Total Environ.* **2019**, *673*, 576–584. [CrossRef]
56. Xu, Y.; Richlen, M.L.; Morton, S.L.; Mak, Y.L.; Chan, L.L.; Tekiau, A.; Anderson, D.M. Distribution, abundance and diversity of Gambierdiscus spp. from a ciguatera-endemic area in Marakei, Republic of Kiribati. *Harmful Algae* **2014**, *34*, 56–68. [CrossRef]
57. Wu, J.J.; Mak, Y.L.; Murphy, M.B.; Lam, J.C.W.; Chan, W.C.; Wang, M.; Chan, L.L.; Lam, P.K.S. Validation of an accelerated solvent extraction liquid chromatography tandem mass spectrometry method for Pacific ciguatoxin-1 in fish flesh and comparison with the mouse neuroblastoma assay. *Anal. Bioanal. Chem.* **2011**, *400*, 3165–3175. [CrossRef] [PubMed]
58. Robertson, A.; Garcia, A.C.; Quintana, H.A.F.; Smith, T.B.; Castillo, B.F., II; Reale-Munroe, K.; Gulli, J.A.; Olsen, D.A.; Hooe-Rollman, J.I.; Jester, E.L.E.; et al. Invasive lionfish (Pterois volitans): A potential human health threat for ciguatera fish poisoning in tropical waters. *Mar. Drugs* **2014**, *12*, 88–97. [CrossRef] [PubMed]
59. Roué, M.; Darius, H.T.; Picot, S.; Ung, A.; Viallon, J.; Gaertner-Mazouni, N.; Sibat, M.; Amzil, Z.; Chinain, M. Evidence of the bioaccumulation of ciguatoxins in giant clams (Tridacna maxima) exposed to *Gambierdiscus* spp. cells. *Harmful Algae* **2016**, *57*, 78–87. [CrossRef] [PubMed]
60. Okumura, M.; Tsuzuki, H.; Tomita, B.-I. A rapid detection method for paralytic shellfish poisoning toxins by cell bioassay. *Toxicon* **2005**, *46*, 93–98. [CrossRef]
61. Mak, Y.L.; Wai, T.-C.; Murphy, M.B.; Chan, W.H.; Wu, J.J.; Lam, J.C.W.; Chan, L.L.; Lam, P.K.S. Pacific ciguatoxins in food web components of coral reef systems in the Republic of Kiribati. *Environ. Sci. Technol.* **2013**, *47*, 14070–14079. [CrossRef]
62. Loeffler, C.R.; Handy, S.M.; Flores Quintana, H.A.; Deeds, J.R. Fish hybridization leads to uncertainty regarding ciguatera fish poisoning risk; confirmation of hybridization and ciguatoxin accumulation with implications for stakeholders. *J. Mar. Sci. Eng.* **2019**, *7*, 105. [CrossRef]
63. Lewis, R.J.; Inserra, M.; Vetter, I.; Holland, W.C.; Hardison, D.R.; Tester, P.A.; Litaker, R.W. Rapid extraction and identification of maitotoxin and ciguatoxin-like toxins from Caribbean and Pacific Gambierdiscus using a new functional bioassay. *PLoS ONE* **2016**, *11*, e0160006. [CrossRef]

64. Pisapia, F.; Holland, W.C.; Hardison, D.R.; Litaker, R.W.; Fraga, S.; Nishimura, T.; Adachi, M.; Nguyen-Ngoc, L.; Séchet, V.; Amzil, Z.; et al. Toxicity screening of 13 Gambierdiscus strains using neuro-2a and erythrocyte lysis bioassays. *Harmful Algae* **2017**, *63*, 173–183. [CrossRef]
65. Litaker, R.W.; Holland, W.C.; Hardison, D.R.; Pisapia, F.; Hess, P.; Kibler, S.R. Ciguatoxicity of *Gambierdiscus* and *Fukuyoa* species from the Caribbean and Gulf of Mexico. *PLoS ONE* **2017**, *12*, e0185776. [CrossRef]
66. Ledreux, A.; Sérandour, A.-L.; Morin, B.; Derick, S.; Lanceleur, R.; Hamlaoui, S.; Furger, C.; Biré, R.; Krys, S.; Fessard, V.; et al. Collaborative study for the detection of toxic compounds in shellfish extracts using cell-based assays. Part II: Application to shellfish extracts spiked with lipophilic marine toxins. *Anal. Bioanal. Chem.* **2012**, *403*, 1995–2007. [CrossRef]
67. Ledreux, A.; Krys, S.; Bernard, C. Suitability of the Neuro-2a cell line for the detection of palytoxin and analogues (neurotoxic phycotoxins). *Toxicon* **2009**, *53*, 300–308. [CrossRef] [PubMed]
68. Laurent, D.; Kerbrat, A.S.; Darius, H.T.; Rossi, F.; Yeeting, B.; Haddad, M.; Golubic, S.; Pauillac, S.; Chinain, M. Ciguatera Shellfish Poisoning (CSP): A new ecotoxicological phenomenon from cyanobacteria to humans via giant clams. In *Food Chains: New Research*; Jensen, M., Muller, D.W., Eds.; Nova Science Publisher Inc.: New York, NY, USA, 2012; pp. 1–44.
69. Kosar, B.N.; West, K.L.; DeFelice, S.V.; Bienfang, P.K. Neurotoxin prevalence from stranded Hawaiian cetaceans. *Glob. Adv. Res. J. Environ. Sci. Toxicol. (GARJEST)* **2015**, *4*, 8–14.
70. Kerbrat, A.S.; Amzil, Z.; Pawlowiez, R.; Golubic, S.; Sibat, M.; Darius, H.T.; Chinain, M.; Laurent, D. First evidence of palytoxin and 42-hydroxy-palytoxin in the marine cyanobacterium Trichodesmium. *Mar. Drugs* **2011**, *9*, 543–560. [CrossRef]
71. Jiang, X.-W.; Li, X.; Lam, P.K.S.; Cheng, S.H.; Schlenk, D.; de Mitcheson, Y.S.; Li, Y.; Gu, J.-D.; Chan, L.L. Proteomic analysis of hepatic tissue of ciguatoxin (CTX) contaminated coral reef fish Cephalopholis argus and moray eel Gymnothorax undulatus. *Harmful Algae* **2012**, *13*, 65–71. [CrossRef]
72. Hardison, D.R.; Holland, W.C.; Darius, H.T.; Chinain, M.; Tester, P.A.; Shea, D.; Bogdanoff, A.K.; Morris, J.A., Jr.; Flores Quintana, H.A.; Loeffler, C.R.; et al. Investigation of ciguatoxins in invasive lionfish from the greater caribbean region: Implications for fishery development. *PLoS ONE* **2018**, *13*, e0198358. [CrossRef] [PubMed]
73. Hardison, D.R.; Holland, W.C.; McCall, J.R.; Bourdelais, A.J.; Baden, D.G.; Darius, H.T.; Chinain, M.; Tester, P.A.; Shea, D.; Flores Quintana, H.A.; et al. Fluorescent Receptor Binding Assay for detecting ciguatoxins in fish. *PLoS ONE* **2016**, *11*, e0153348. [CrossRef] [PubMed]
74. Hossen, V.; Soliño, L.; Leroy, P.; David, E.; Velge, P.; Dragacci, S.; Krys, S.; Flores Quintana, H.; Diogène, J. Contribution to the risk characterization of ciguatoxins: LOAEL estimated from eight ciguatera fish poisoning events in Guadeloupe (French West Indies). *Environ. Res.* **2015**, *143*, 100–108. [CrossRef]
75. Hokama, Y.; Chun, K.E.; Campora, C.E.; Higa, N.; Suma, C.; Hamajima, A.; Isobe, M. Biological activity of the functional epitope of ciguatoxin fragment AB on the neuroblastoma sodium channel in tissue culture. *J. Clin. Lab. Anal.* **2006**, *20*, 126–132. [CrossRef]
76. Dechraoui, M.Y.; Naar, J.; Pauillac, S.; Legrand, A.M. Ciguatoxins and brevetoxins, neurotoxic polyether compounds active on sodium channels. *Toxicon* **1999**, *37*, 125–143. [CrossRef]
77. Darius, H.T.; Roué, M.; Sibat, M.; Viallon, J.; Gatti, C.M.; Vandersea, M.W.; Tester, P.A.; Litaker, R.W.; Amzil, Z.; Hess, P.; et al. Tectus niloticus (Tegulidae, Gastropod) as a novel vector of Ciguatera Poisoning: Detection of Pacific ciguatoxins in toxic samples from Nuku Hiva Island (French Polynesia). *Toxins* **2018**, *10*, 2. [CrossRef]
78. Darius, H.T.; Roué, M.; Sibat, M.; Viallon, J.; Gatti, C.M.I.; Vandersea, M.W.; Tester, P.A.; Litaker, R.W.; Amzil, Z.; Hess, P.; et al. Toxicological investigations on the sea urchin Tripneustes gratilla (Toxopneustidae, Echinoid) from Anaho Bay (Nuku Hiva, French Polynesia): Evidence for the presence of Pacific ciguatoxins. *Mar. Drugs* **2018**, *16*, 122. [CrossRef] [PubMed]
79. Dai, X.; Mak, Y.L.; Lu, C.-K.; Mei, H.-H.; Wu, J.J.; Lee, W.H.; Chan, L.L.; Lim, P.T.; Mustapa, N.I.; Lim, H.C.; et al. Taxonomic assignment of the benthic toxigenic dinoflagellate Gambierdiscus sp. type 6 as Gambierdiscus balechii (Dinophyceae), including its distribution and ciguatoxicity. *Harmful Algae* **2017**, *67*, 107–1[illegible]. [CrossRef] [PubMed]
80. Chan, W.H.; Mak, Y.L.; Wu, J.J.; Jin, L.; Sit, W.H.; Lam, J.C.W.; Sadovy de Mitcheson, Y.; Chan, L.L.; Lam, P.K.S.; Murphy, M.B. Spatial distribution of ciguateric fish in the Republic of Kiribati. *Chemosphere* **2011**, *84*, 117–1[illegible]. [CrossRef] [PubMed]

81. Caillaud, A.; Eixarch, H.; de la Iglesia, P.; Rodriguez, M.; Dominguez, L.; Andree, K.B.; Diogène, J. Towards the standardisation of the neuroblastoma (neuro-2a) cell-based assay for ciguatoxin-like toxicity detection in fish: Application to fish caught in the Canary Islands. *Food Addit. Contam. Part A* **2012**, *29*, 1000–1010. [CrossRef] [PubMed]
82. Caillaud, A.; Cañete, E.; de la Iglesia, P.; Giménez, G.; Diogène, J. Cell-based assay coupled with chromatographic fractioning: A strategy for marine toxins detection in natural samples. *Toxicol. In Vitro* **2009**, *23*, 1591–1596. [CrossRef] [PubMed]
83. Canete, E.; Diogene, J. Comparative study of the use of neuroblastoma cells (Neuro-2a) and neuroblastomaxglioma hybrid cells (NG108-15) for the toxic effect quantification of marine toxins. *Toxicon* **2008**, *52*, 541–550. [CrossRef] [PubMed]
84. Campora, C.E.; Dierking, J.; Tamaru, C.S.; Hokama, Y.; Vincent, D. Detection of ciguatoxin in fish tissue using sandwich Elisa and neuroblastoma cell bioassay. *J. Clin. Lab. Anal.* **2008**, *22*, 246–253. [CrossRef]
85. Bravo, J.; Suarez, F.C.; Ramirez, A.S.; Acosta, F. Ciguatera, an emerging human poisoning in Europe. *J. Aquac. Mar. Biol.* **2015**, *3*, 00053.
86. Bienfang, P.; Oben, B.; DeFelice, S.; Moeller, P.; Huncik, K.; Oben, P.; Toonen, R.; Daly-Engel, T.; Bowen, B. Ciguatera: The detection of neurotoxins in carnivorous reef fish from the coast of Cameroon, West Africa. *Afr. J. Mar. Sci.* **2008**, *30*, 533–540. [CrossRef]
87. Bodero, M.; Bovee, T.F.H.; Wang, S.; Hoogenboom, R.L.A.P.; Klijnstra, M.D.; Portier, L.; Hendriksen, P.J.M.; Gerssen, A. Screening for the presence of lipophilic marine biotoxins in shellfish samples using the neuro-2a bioassay. *Food Addit. Contam. Part A* **2018**, *35*, 351–365. [CrossRef] [PubMed]
88. Bodero, M.; Gerssen, A.; Portier, L.; Klijnstra, M.D.; Hoogenboom, R.L.A.P.; Guzmán, L.; Hendriksen, P.J.M.; Bovee, T.F.H. A strategy to replace the mouse bioassay for detecting and identifying lipophilic marine biotoxins by combining the Neuro-2a bioassay and LC-MS/MS analysis. *Mar. Drugs* **2018**, *16*, 501. [CrossRef] [PubMed]
89. O'Toole, A.C.; Dechraoui Bottein, M.-Y.; Danylchuk, A.J.; Ramsdell, J.S.; Cooke, S.J. Linking ciguatera poisoning to spatial ecology of fish: A novel approach to examining the distribution of biotoxin levels in the great barracuda by combining non-lethal blood sampling and biotelemetry. *Sci. Total Environ.* **2012**, *427–428*, 98–105. [CrossRef] [PubMed]
90. Bottein Dechraoui, M.-Y.; Wang, Z.; Ramsdell, J.S. Optimization of ciguatoxin extraction method from blood for Pacific ciguatoxin (P-CTX-1). *Toxicon* **2007**, *49*, 100–105. [CrossRef] [PubMed]
91. Reverté, L.; Toldrà, A.; Andree, K.B.; Fraga, S.; Falco, G.; Campàs, M.; Diogène, J. Assessment of cytotoxicity in ten strains of *Gambierdiscus australes* from Macaronesian Islands by neuro-2a cell-based assays. *J. Appl. Phycol.* **2018**, *30*, 2447–2461. [CrossRef]
92. Bottein Dechraoui, M.-Y.; Wang, Z.; Ramsdell, J.S. Intrinsic potency of synthetically prepared brevetoxin cysteine metabolites BTX-B2 and desoxyBTX-B2. *Toxicon* **2007**, *50*, 825–834. [CrossRef] [PubMed]
93. Catania, D.; Richlen, M.L.; Mak, Y.L.; Morton, S.L.; Laban, E.H.; Xu, Y.; Anderson, D.M.; Chan, L.L.; Berumen, M.L. The prevalence of benthic dinoflagellates associated with ciguatera fish poisoning in the central Red Sea. *Harmful Algae* **2017**, *68*, 206–216. [CrossRef]
94. Nicolas, J.A.Y. Innovative Mode of Action Based in Vitro Assays for Detection of Marine Neurotoxins. Ph.D. Thesis, Wageningen University, Wageningen, The Netherlands, 2015.
95. Sanchez-Henao, A.; García-Álvarez, N.; Silva Sergent, F.; Estévez, P.; Gago-Martínez, A.; Martín, F.; Ramos-Sosa, M.; Fernández, A.; Diogène, J.; Real, F. Presence of CTXs in moray eels and dusky groupers in the marine environment of the Canary Islands. *Aquat. Toxicol.* **2020**, *221*, 105427. [CrossRef]
96. Dechraoui-Bottein, M.-Y. Etude du Mode d'Action des Ciguatoxines, Biotoxines Marines Responsable de la Ciguatéra: Comparaison aux Brévétoxines et Application à la Détection des Poissons Toxiques. Ph.D. Thesis, Université Française du Pacifique, Tahiti, French Polynesia, 1999.
97. Perreault, F.; Matias, M.S.; Melegari, S.P.; de Carvalho Pinto, C.R.S.; Creppy, E.E.; Popovic, R.; Matias, W.G. Investigation of animal and algal bioassays for reliable saxitoxin ecotoxicity and cytotoxicity risk evaluation. *Ecotoxicol. Environ. Saf.* **2011**, *74*, 1021–1026. [CrossRef]
98. McCall, J.R.; Elliott, E.A.; Bourdelais, A.J. A new cytotoxicity assay for brevetoxins using fluorescence microscopy. *Mar. Drugs* **2014**, *12*, 4868–4882. [CrossRef]
99. Suzuki, T.; Ha, D.V.; Uesugi, A.; Uchida, H. Analytical challenges to ciguatoxins. *Curr. Opin. Food Sci.* **2017**, *18*, 37–42. [CrossRef]

100. Estevez, P.; Castro, D.; Manuel Leao, J.; Yasumoto, T.; Dickey, R.; Gago-Martinez, A. Implementation of liquid chromatography tandem mass spectrometry for the analysis of ciguatera fish poisoning in contaminated fish samples from Atlantic coasts. *Food Chem.* **2019**, *280*, 8–14. [CrossRef]

101. González, O.; Blanco, M.E.; Iriarte, G.; Bartolomé, L.; Maguregui, M.I.; Alonso, R.M. Bioanalytical chromatographic method validation according to current regulations, with a special focus on the non-well defined parameters limit of quantification, robustness and matrix effect. *J. Chromatogr. A* **2014**, *1353*, 10–27. [CrossRef] [PubMed]

102. Rossini, G.; Hartung, T. Towards tailored assays for cell-based approaches to toxicity testing. *Altern. Anim. Exp. (ALTEX)* **2012**, *29*, 359–372.

103. Diogène, J.; Reverté, L.; Rambla-Alegre, M.; del Río, V.; de la Iglesia, P.; Campàs, M.; Palacios, O.; Flores, C.; Caixach, J.; Ralijaona, C.; et al. Identification of ciguatoxins in a shark involved in a fatal food poisoning in the Indian Ocean. *Sci. Rep.* **2017**, *7*, 8240. [CrossRef] [PubMed]

104. Kerrigan, B.A. Post-settlement growth and body composition in relation to food availability in a juvenile tropical reef fish. *Mar. Ecol. Prog. Ser. Oldend.* **1994**, *111*, 7–15. [CrossRef]

105. Harwood, D.T.; Murray, S.; Boundy, M.J. Chapter Three–Sample Preparation Prior to Marine Toxin Analysis. In *Comprehensive Analytical Chemistry*; Diogène, J., Campàs, M., Eds.; Elsevier: Amsterdam, The Netherlands, 2017; Volume 78, pp. 89–136.

106. Abraham, A.; Jester, E.L.E.; Granade, H.R.; Plakas, S.M.; Dickey, R.W. Caribbean ciguatoxin profile in raw and cooked fish implicated in ciguatera. *Food Chem.* **2012**, *131*, 192–198. [CrossRef]

107. Bienfang, P.; DeFelice, S.; Dowling, A. Quantitative evaluation of commercially available test kit for ciguatera in fish. *Food Nutr. Sci.* **2011**, *2*, 594. [CrossRef]

108. Coccini, T.; Caloni, F.; De Simone, U. Human neuronal cell based assay: A new in vitro model for toxicity evaluation of ciguatoxin. *Environ. Toxicol. Pharmacol.* **2017**, *52*, 200–213. [CrossRef]

109. Díaz-Asencio, L.; Clausing, R.J.; Vandersea, M.; Chamero-Lago, D.; Gómez-Batista, M.; Hernández-Albernas, J.I.; Chomérat, N.; Rojas-Abrahantes, G.; Litaker, R.W.; Tester, P.; et al. Ciguatoxin occurrence in food-web components of a cuban coral reef ecosystem: Risk-assessment implications. *Toxins* **2019**, *11*, 722. [CrossRef]

110. Estevez, P.; Castro, D.; Pequeño-Valtierra, A.; Leao, J.M.; Vilariño, O.; Diogène, J.; Gago-Martínez, A. An attempt to characterize the ciguatoxin profile in Seriola fasciata causing ciguatera fish poisoning in Macaronesia. *Toxins* **2019**, *11*, 221. [CrossRef]

111. Manger, R.L.; Leja, L.S.; Lee, S.Y.; Hungerford, J.M.; Kirkpatrick, M.A.; Yasumoto, T.; Wekell, M.M. Detection of paralytic shellfish poison by rapid cell bioassay: Antagonism of voltage-gated sodium channel active toxins in vitro. *J. AOAC Int.* **2003**, *86*, 540–543. [CrossRef] [PubMed]

112. Silva, M.; Pratheepa, V.; Botana, L.; Vasconcelos, V. Emergent toxins in north Atlantic temperate waters: A challenge for monitoring programs and legislation. *Toxins* **2015**, *7*, 859. [CrossRef] [PubMed]

113. Darius, H.T.; Ponton, D.; Revel, T.; Cruchet, P.; Ung, A.; Tchou Fouc, M.; Chinain, M. Ciguatera risk assessment in two toxic sites of French Polynesia using the receptor-binding assay. *Toxicon* **2007**, *50*, 612–626. [CrossRef] [PubMed]

114. Konoki, K.; Baden, D.G.; Scheuer, T.; Catterall, W.A. Molecular determinants of brevetoxin binding to voltage-gated sodium channels. *Toxins* **2019**, *11*, 513. [CrossRef]

115. Schlumberger, S.; Mattei, C.; Molgó, J.; Benoit, E. Dual action of a dinoflagellate-derived precursor of Pacific ciguatoxins (P-CTX-4B) on voltage-dependent K+ and Na+ channels of single myelinated axons. *Toxicon* **2010**, *56*, 768–775. [CrossRef]

116. Mattei, C.; Legros, C. The voltage-gated sodium channel: A major target of marine neurotoxins. *Toxicon* **2014**, *91*, 84–95. [CrossRef]

117. Mattei, C.; Molgó, J.; Benoit, E. Involvement of both sodium influx and potassium efflux in ciguatoxin-induced nodal swelling of frog myelinated axons. *Neuropharmacology* **2014**, *85*, 417–426. [CrossRef]

118. EFSA. Scientific opinion on marine biotoxins in shellfish—Emerging toxins: Ciguatoxin Group. *EFSA J.* **2010**, *1627*, 1–38.

119. Food and Drug Administration (FDA). *Fish and Fishery Products Hazards and Control Guidance*; FDA: Washington, DC, USA, 2019; pp. 1–19.

120. Sibat, M.; Herrenknecht, C.; Darius, H.T.; Roué, M.; Chinain, M.; Hess, P. Detection of Pacific ciguatoxins using liquid chromatography coupled to either low or high resolution mass spectrometry (LC-MS/MS). *J. Chromatogr. A* **2018**, *1571*, 16–28. [CrossRef]
121. Costa, P.R.; Estevez, P.; Castro, D.; Soliño, L.; Gouveia, N.; Santos, C.; Rodrigues, S.M.; Leao, J.M.; Gago-Martínez, A. New insights into the occurrence and toxin profile of ciguatoxins in Selvagens Islands (Madeira, Portugal). *Toxins* **2018**, *10*, 524. [CrossRef]
122. Murray, J.S.; Boundy, M.J.; Selwood, A.I.; Harwood, D.T. Development of an LC–MS/MS method to simultaneously monitor maitotoxins and selected ciguatoxins in algal cultures and P-CTX-1B in fish. *Harmful Algae* **2018**, *80*, 80–87. [CrossRef] [PubMed]
123. Yogi, K.; Sakugawa, S.; Oshiro, N.; Ikehara, T.; Sugiyama, K.; Yasumoto, T. Determination of toxins involved in Ciguatera Fish Poisoning in the Pacific by LC/MS. *J. AOAC Int.* **2014**, *97*, 398–402. [CrossRef] [PubMed]
124. Ikehara, T.; Kuniyoshi, K.; Oshiro, N.; Yasumoto, T. Biooxidation of ciguatoxins leads to species-specific toxin profiles. *Toxins* **2017**, *9*, 205. [CrossRef]
125. Yogi, K.; Oshiro, N.; Inafuku, Y.; Hirama, M.; Yasumoto, T. Detailed LC-MS/MS analysis of ciguatoxins revealing distinct regional and species characteristics in fish and causative alga from the Pacific. *Anal. Chem.* **2011**, *83*, 8886–8891. [CrossRef]
126. Mak, Y.L.; Wu, J.J.; Chan, W.H.; Murphy, M.B.; Lam, J.C.W.; Chan, L.L.; Lam, P.K.S. Simultaneous quantification of Pacific ciguatoxins in fish blood using liquid chromatography–tandem mass spectrometry. *Anal. Bioanal. Chem.* **2013**, *405*, 3331–3340. [CrossRef]
127. Díaz-Asencio, L.; Clausing, R.J.; Rañada, M.L.; Alonso-Hernández, C.M.; Dechraoui Bottein, M.-Y. A radioligand receptor binding assay for ciguatoxin monitoring in environmental samples: Method development and determination of quality control criteria. *J. Environ. Radioact.* **2018**, *192*, 289–294. [CrossRef]
128. Van Dolah, F.M.; Fire, S.E.; Leighfield, T.A.; Mikulski, C.M.; Doucette, G.J. Determination of paralytic shellfish toxins in shellfish by receptor binding assay: Collaborative study. *J. AOAC Int. Food Chem. Contam.* **2012**, *95*, 795–812. [CrossRef]
129. Van Dolah, F.M.; Leighfield, T.A.; Doucette, G.J. Single-laboratory validation of the microplate receptor binding assay for paralytic shellfish toxins in shellfish. *J. AOAC Int. Food Chem. Contam.* **2009**, *92*, 1705–1713.
130. Stanbridge, E. Mycoplasma and cell cultures. *Bacteriol. Rev.* **1971**, *35*, 206–227. [CrossRef]
131. Nikfarjam, L.; Farzaneh, P. Prevention and detection of mycoplasma conatamination in cell culture. *Cell J.* **2012**, *13*, 203–212.
132. Chang-Liu, C.-M.; Woloschak, G.E. Effect of passage number on cellular response to DNA-damaging agents: Cell survival and gene expression. *Cancer Lett.* **1997**, *113*, 77–86. [CrossRef]
133. Wenger, S.L.; Senft, J.R.; Sargent, L.M.; Bamezai, R.; Bairwa, N.; Grant, S.G. Comparison of established cell lines at different passages by karyotype and comparative genomic hybridization. *Biosci. Rep.* **2005**, *24*, 631–639. [CrossRef]
134. Loeffler, C.R.; Bodi, D.; Preib-Weigert, A. Neuroblastoma cell-based assay optimization for harmonized marine biotoxin detection: Method for reducing oversensitivity to ouabain (O) and veratridine (V). In Proceedings of the 18th International Conference on Harmful Algae, Nantes, France, 21–26 October 2018; p. 413.
135. Neves, R.A.F.; Pardal, M.A.; Nascimento, S.M.; Oliveira, P.J.; Rodrigues, E.T. Screening-level evaluation of marine benthic dinoflagellates toxicity using mammalian cell lines. *Ecotoxicol. Environ. Saf.* **2020**, *195*, 110465. [CrossRef]
136. Longo, S.; Sibat, M.; Viallon, J.; Darius, H.T.; Hess, P.; Chinain, M. Intraspecific variability in the toxin production and toxin profiles of in vitro cultures of Gambierdiscus polynesiensis (Dinophyceae) from French Polynesia. *Toxins* **2019**, *11*, 735. [CrossRef]
137. Roué, M.; Darius, H.T.; Ung, A.; Viallon, J.; Sibat, M.; Hess, P.; Amzil, Z.; Chinain, M. Tissue distribution and elimination of ciguatoxins in Tridacna maxima (Tridacnidae, Bivalvia) fed Gambierdiscus polynesiensis. *Toxins* **2018**, *10*, 189. [CrossRef]
138. Roué, M.; Darius, H.T.; Gatti, C.; Chinain, M.; Harwood, D.T. Solid phase adsorption toxin tracking (SPATT) technology for field monitoring of Gambierdiscus toxins with passive samplers. *Harmful Algal News* **2018**, *60*, 8–9.
139. Roué, M.; Darius, H.T.; Viallon, J.; Ung, A.; Gatti, C.; Harwood, D.T.; Chinain, M. Application of solid phase adsorption toxin tracking (SPATT) devices for the field detection of Gambierdiscus toxins. *Harmful Algae* **2018**, *71*, 40–49. [CrossRef]

140. Roué, M.; Darius, H.; Chinain, M. Solid phase adsorption toxin tracking (spatt) technology for the monitoring of aquatic toxins: A review. *Toxins* **2018**, *10*, 167. [CrossRef]
141. Chinain, M.; Faust, M.; Pauillac, S. Morphology and molecular analyses of three toxic species of Gambierdiscus (dinophyceae): *G. pacificus*, sp. nov., *G. australes*, sp. nov., and *G. polynesiensis*, sp. nov. *J. Phycol.* **1999**, *35*, 1282–1296. [CrossRef]

toxins

MDPI

Article

Development and Characterization of Monoclonal Antibodies for the Mycotoxin Citreoviridin

Chris M. Maragos [1,*], Yosuke Uchiyama [2], Naoki Kobayashi [2], Fumichika Kominato [3] and Yoshiko Sugita-Konishi [2]

[1] Mycotoxin Prevention and Applied Microbiology Research Unit, Agricultural Research Service, U.S. Department of Agriculture, Peoria, IL 61604, USA

[2] Department of Food and Life Science, Graduate School of Life and Environmental Sciences, Azabu University, Sagamihara, Kanagawa 252-5201, Japan; de1701@azabu-u.ac.jp (Y.U.); n-kobayashi@azabu-u.ac.jp (N.K.); y-konishi@azabu-u.ac.jp (Y.S.-K.)

[3] Shinsei Chemical Company Ltd., Ibaraki, Osaka 567-0085, Japan; kominato@schem.jp

* Correspondence: chris.maragos@usda.gov

Received: 10 October 2019; Accepted: 28 October 2019; Published: 30 October 2019

Abstract: Citreoviridin (CTV) in an inhibitor of mitochondrial ATPase that has been isolated from molded yellow rice and linked to the human disease Shoshin-kakke (acute cardiac beriberi). The disease results from a deficiency of thiamine, however, purified CTV can reproduce the symptoms in experimental animals. The link between CTV and Shoshin-kakke has been difficult to resolve, in part because cases of the disease are rare. In addition to rice, CTV has been found in maize, pecan nuts, and wheat products. A method to screen for CTV and its geometric isomer, iso-CTV, in commodities was developed, based upon the isolation of two novel monoclonal antibodies (mAb). In an antigen-immobilized competitive enzyme-linked immunosorbent assay format (CI-ELISA), the observed IC_{50}s for CTV were 11 ng/mL and 18 ng/mL (mAbs 2-2 and 2-4, respectively). The assays were relatively tolerant to methanol and acetonitrile, which allowed their application to the detection of CTV in spiked polished white rice. For quantification, a standard mixture of CTV and iso-CTV was used, along with matrix matched calibration. The dynamic range of the ELISA using mAb 2-4 was equivalent to 0.23 to 2.22 mg/kg in rice. Recoveries over the range of 0.36 to 7.23 mg/kg averaged 97 ± 10%. The results suggest that the mAb 2-4-based immunoassay can be applied to the screening of white rice for CTV. Both mAbs were also observed to significantly enhance the fluorescence of the toxin.

Keywords: citreoviridin; antibody; immunoassay; rice

Key Contribution: Two monoclonal antibodies were produced and used to develop a screening immunoassay for detecting the mycotoxin citreoviridin at relevant levels in white rice.

1. Introduction

In Japan in the years before World War I and continuing through the 1920s, there were human illnesses associated with consumption of moldy, yellow, rice. The illness, classified at the time as Shoshin-kakke (acute cardiac beriberi) decreased in incidence significantly around 1910, a fact attributed to increased inspection of rice by Japanese authorities [1]. The disease was related to beriberi, now known to be caused by thiamine deficiency. By 1930 Shoshin-kakke had almost completely disappeared from Japan [1]. Subsequent investigations in Japan led to the differentiation of the broad category of yellow-colored rice into five groups, of which four are caused by Penicillium spoilage and one by *Eurotium amstelodami* [2]. The four caused by Penicillia are each associated with a different species of fungus and a different causative agent. The type of yellow rice known as Ou-hen-mai is infested with *Penicillium citreonigrum* and has been associated with Shoshin-kakke. In 1964 the structure of the mycotoxin believed to be the causative agent (citreoviridin, CTV) was reported [3]. Another type of yellow rice is Citrinum yellow rice (Citrinum ou-hen-mai) where the causative fungus is *P. citrinum* and the associated mycotoxin is citrinin. In 2006–2008 an outbreak of beriberi occurred in the Maranhão state of Brazil. Despite the presence of a few samples contaminated with CTV, the cases appear to have been predominantly a result of thiamine deficiency, as many were reversed following administration of thiamine [4–6]. An excellent summary of the history of yellow rice and the classification of rice infested with fungi was provided by Kushiro [2].

The connection of Shoshin-kakke to moldy rice has been confounded by the multiple types of "yellow rice", the low incidence of the disease in modern times, and the extent to which thiamine deficiency is required to produce symptoms. In research conducted in the 1960s and early 1970s, crude alcohol extracts of moldy rice were tested in 14 vertebrate species [1,7]. Symptoms included paralysis of the legs, vomiting, convulsions, and respiratory arrest [7]. Purified CTV given to mice, cats, and dogs reproduced these symptoms, with an LD_{50} of 20 mg/kg in mice [1]. Purified CTV given to mice and rats was lethal, with LD_{50}s ranging from 3.6 to 11 mg/kg [7]. Reproducing symptoms of Shoshin-kakke with purified CTV was important for distinguishing intoxication due to consumption of yellow rice, from disease caused directly by thiamine deficiency. As such, Shoshin-kakke is considered to be a mycotoxicosis [2].

On the molecular level, CTV inhibits the mitochondrial adenosine triphosphatase (ATPase) [8,9]. When given to rats CTV altered the pattern of transketolase (EC 2.2.1.1) in liver, and in vitro experiments suggested an anti-thiamine effect of the toxin [10]. A mechanism that involves the exacerbation or causation of thiamine deficiency would be consistent with the involvement of CTV in Shoshin-kakke. Recently, the toxicokinetics of CTV was determined in swine [11]. Results suggested that following oral exposure, CTV was readily absorbed by swine, and slowly metabolized, with a half-life of 21 h [11].

CTV (Figure 1) is produced by a variety of fungi including *P. citreonigrum*, *Aspergillus terreus*, and *Eupenicillium ochrosalmoneum* (actual identity *P. ochrosalmoneum*) [12–14]. The fungi that produce CTV were summarized recently by Peterson et al. [15]. While it is known primarily for its association with rice in Asia, CTV has been found in maize and pecan nuts in the United States [16,17]. It has also been found in rice and wheat-based products in Brazil [18] and in grain dust in Belgium [19].

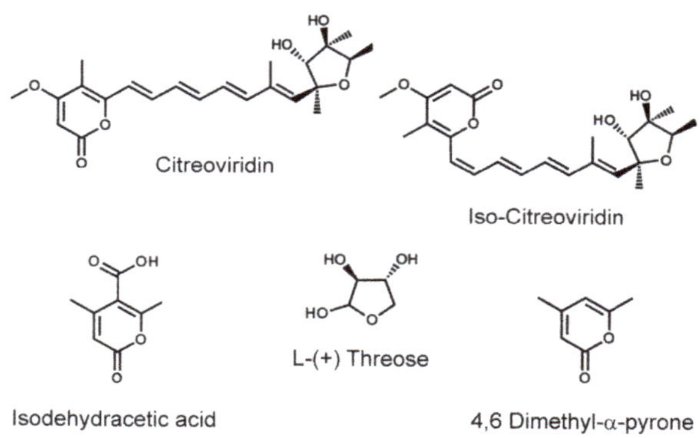

Figure 1. Structure of citreoviridin (CTV) and related compounds.

The structure of CTV was first determined in 1964 [3]. Many of the physical properties of CTV were reported by Ueno and Ueno [7] and were summarized by Cole and Cox [20]. Purified CTV is bright yellow and, in methanol, has absorption maxima in the ultraviolet at 383 nm (ε 44,925), 294 nm (ε 24,725), 285 nm (ε 23,343), 238 nm (ε 10,383), and 203 nm (ε 15,388) [21]. Similar absorption maxima, but lower molar absorptivity coefficients were reported more recently, also in methanol: 387 nm (ε 31,590), 294 nm (ε 21,960), and 285 nm (ε 20,060) [13]. The latter also reported Fourier transform infrared (FTIR), ^{1}H-, and ^{13}C-nuclear magnetic resonance (NMR) spectra of CTV. Early reports established that CTV was fluorescent [7,22].

CTV has been extracted from commodities and fungal cultures with a variety of solvents, including ethanol (used in the early toxicity testing), chloroform [13,17], dichloromethane [21] and aqueous methanol [23–25]. Before 1988 most analytical methods for detecting CTV relied upon thin-layer chromatography, however, in that year a normal phase liquid-chromatography method with fluorescence detection (LC-FLD) was reported [21]. With a mobile phase of ethyl acetate-hexane (3+1) the fluorescence was detected at 480 nm using an excitation of 388 nm. More recently, normal phase LC-FLD was used to detect CTV in rice [4]. CTV has also been measured using reverse phase LC with photodiode array detection (LC-PDA) [13]. Several laboratories have reported reverse phase LC-tandem mass spectrometry (LC-MS/MS) methods [13]. The molecular ion in positive mode was observed at *m/z* 403.2, with the main fragments at *m/z* 315 and 139 [13,18]. Detailed mass spectra were reported by Rebuffat et al. [26].

The total synthesis of CTV gave a product that upon exposure to ambient light yielded a mixture of two major components. This led to the realization authentic CTV, which exists as the all-trans form, undergoes photoisomerization, with the product termed "iso-CTV" [27]. The ratio of CTV:iso-CTV has been reported as 7:3 [28] and 3:2 [4,27]. Handling the purified CTV only under red light minimized the isomerization, however handling the toxin under typical laboratory ambient light resulted in the mixture reaching a photostable state within only 1 to 9 h [27]. Even when stored under frozen conditions and protected from light, CTV has been reported to isomerize [13]. As such it should be considered that, under typical laboratory conditions, preparations of "citreoviridin" will likely exist as mixtures that at equilibrium have ratios of CTV:iso-CTV ranging from 1.5:1 to 2.3:1.

Antibodies against CTV have been reported previously. These included polyclonal antibodies [23], a monoclonal antibody (mAb) [24], and a single chain variable fragment (scFv) antibody [25]. While these represented important efforts, the sensitivities of the immunoassays allowed for improvement. In this report we undertook to develop monoclonal antibodies for the detection of CTV, to improve immunoassays for CTV, and to apply such immunoassays to spiked white rice. In an attempt to yield improved antibodies, CTV-protein conjugates were prepared using a different approach from that described previously. Following the development of the mAbs, an interesting effect of the antibodies on the fluorescence of CTV was observed and was likewise studied.

2. Results and Discussion

2.1. Antibody Development, and Tolerance to Solvents

Because of its low molecular weight, CTV was first conjugated to bovine serum albumin (BSA) prior to immunization of mice. To facilitate the identification of CTV-binding antibodies, an ovalbumin (OVA) conjugate was also prepared and used as the immobilized antigen in an indirect competitive enzyme-linked immunosorbent assay (CI-ELISA). Previous attempts to produce CTV antibodies either oxidized the hydroxy groups on CTV and then linked them to proteins [23], or formed hemisuccinate derivatives that were then linked to the proteins [24]. Both routes involved linkage through the hydroxy groups of CTV. In our attempt, the linkage was also made through the hydroxy groups, but using a carbodiimide-based chemistry that did not involve oxidation of the hydroxy groups first or require the introduction of a long linker group. When reacted with hydroxy groups, the reagent used, 1, 1′-carbonyldiimidazole, forms a carbamate linkage with primary amines of the proteins, adding only one carbon to the length of the linkage [29]. The BSA conjugate (CTV-BSA3) was administered to 10 mice and their sera was evaluated for binding to the OVA conjugate (CTV-OVA2) and for response to free CTV. Two of the sera showed binding to the immobilized antigen (CTV-OVA2) and were selected for sacrifice and splenocyte fusions. No viable products were collected from the first fusion, while the second fusion yielded only five products. Of the five, three were able to bind free CTV. Attempts to subclone these three yielded two stable cell lines designated herein as "2-2" and "2-4". Antibodies produced by these two cell lines were evaluated in a CI-ELISA format for their sensitivity towards free CTV, cross-reactivity towards small molecules with similar structures, tolerance to solvents, and ability to detect CTV in spiked rice.

The antibodies were relatively sensitive to CTV in buffer (0.1% OVA-phosphate buffered saline (PBS)), with IC_{50}s in the range of 11 to 18 ng/mL. The analytical standard used to determine the IC_{50}s contained predominantly CTV, with a small proportion of iso-CTV (7%, Figure 2A).

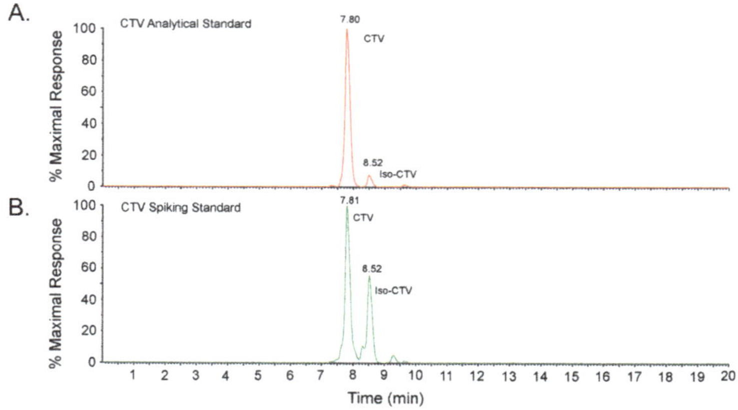

Figure 2. Liquid chromatography-tandem mass spectrometry (LC-MS) traces of protonated CTV and iso-CTV (m/z = 403.2). (**A**) The analytical standard used for establishing the response of the competitive enzyme-linked immunosorbent assays (CI-ELISAs) under various solvent conditions (93% CTV, 7% iso-CTV). (**B**) The standard mixture used for examining cross-reactivity to iso-CTV and for spiking rice samples (64% CTV, 36% iso-CTV).

Three compounds having structures that resembled portions of CTV were examined for cross reactivity, including 4,6-dimethyl-α-pyrone (DMP), iso-dehydracetic acid (IDHA), and L-(+)-threose (Figure 1). DMP and IDHA have some similarity to the lactone portion of CTV. L-(+)-threose has some similarity to portions of the substituted furan of CTV. Neither IDHA nor L-(+)-threose were recognized by either of the mAbs, even at levels as high as 500 μg/mL, indicating a cross-reactivity of less than 0.004%. There was a slight inhibition with DMP at 500 μg/mL, indicating a slight cross reaction on the order of 0.04%. Obtaining IC_{50} data for DMP would have required concentrations above the solubility limit for this compound in the test buffer, hence the upper limit on the cross-reactivity. Of course, the most interesting compound to test for cross-reactivity was iso-CTV. This material was not available commercially as an analytical standard, likely because of the photoconversion between iso-CTV and CTV. However, some insights were obtained by comparing responses of standard curves prepared with a small proportion of iso-CTV (7%, Figure 2A) to those prepared with a higher proportion of iso-CTV (36%) (Figure 2B). The comparison was made by accounting for the total of the CTV and iso-CTV present. As indicated in Figures S1 and S2, when the total of CTV and iso-CTV was accounted for, the calibration curves obtained with both mixtures were nearly superimposable, with nearly identical IC_{50}s. Put another way, if the calibration curves were not corrected for the presence of the iso-CTV, the preparation in Figure 2B (high iso-CTV) gave a much better ELISA response than the preparation in Figure 2A (low iso-CTV). The strong similarities between the two curves, when iso-CTV was accounted for, provided indirect evidence that the antibodies also recognize iso-CTV. Unfortunately, the absence of a relatively pure iso-CTV standard prevented us from establishing this directly.

Citreoviridin is relatively hydrophobic, with a predicted Log P of 3.04, compared to 2.68 for toluene and 3.94 for n-hexane [30]. Because of this, the conditions to extract CTV and iso-CTV from commodities use organic solvents or aqueous mixtures of organic solvents. For this reason, it was important to determine the impact of solvent concentration upon the CI-ELISAs of the two antibodies. Both antibodies demonstrated good tolerance to methanol and poorer tolerance to acetonitrile (Table 1). There was not a significant impact of methanol until the concentration used to prepare the standards reached 30% (*v*/*v*). At that point, the solubility of the OVA in the test buffer began to fail, and the variability amongst replicates worsened. The proteins began to precipitate at a lower concentration with acetonitrile (20%), which limited the concentration tested here to 15%. For the best results the methanol concentration should be kept at or below 20% and the acetonitrile concentration at or below 10%.

Table 1. Effects of methanol and acetonitrile on the CI-ELISAs based upon mAbs 2-2 and 2-4.

Solvent	Concentration [a]	IC_{50} for CTV (ng/mL)		Replicates [b]
		mAb 2-2	**mAb 2-4**	
Buffer [c]	-	11.1 ± 3.2	17.9 ± 5.6	8
Methanol	10%	14.8 ± 1.2	20.4 ± 0.5	4
	15%	13.7 ± 1.2	21.7 ± 4.6	6
	20%	15.2 ± 1.6	21.0 ± 1.1	4
	30%	24.9 ± 14.6	36.9 ± 12.5	4
Acetonitrile	5%	13.3 ± 2.1	20.7 ± 1.1	4
	10%	18.7 ± 3.3	29.2 ± 4.6	4
	15%	29.6 ± 2.5	43.4 ± 4.2	4

[a] Concentration of solvent in which the CTV standards were prepared (percentage, *v*/*v*). The concentration of solvent in the competition mixture is half of this value. [b] Number of replicate plates used in the statistic. [c] 0.1% ovalbumin-phosphate buffered saline (OVA-PBS).

As noted previously, the mAbs 2-2 and 2-4 are not the first antibodies that have been applied to immunoassays for CTV detection. The development of mouse polyclonal antibodies (pAb), mouse mAbs, and a single chain variable fragment (scFv) antibody were reported by researchers at the Fujian Agricuture and Forestry University [23–25]. These are the only known reports of CTV antibodies, and the response of the mAbs 2-2 and 2-4 compared well to them, with lower IC_{50}s and working ranges (Table 2).

Table 2. Immunoassays for CTV.

Type of Antibody	IC_{50} for CTV (ng/mL)	Dynamic Range (IC_{20} to IC_{80}) (ng/mL)	Citation
pAb	560	Not specified	[23]
mAb	161	11–2370	[24]
scFv	120	25–562	[25]
mAb	11	3–29	this work, mAb 2-2
mAb	18	5–42	this work, mAb 2-4

2.2. *Application of mAb 2-4 Competitive Enzyme-Linked Immunosorbent Assay (CI-ELISA) to Spiked Rice*

CTV has been extracted from commodities using a wide range of solvents, from pure dichloromethane to 5% methanol [13,17,23–25]. To ensure compatibility with the ELISAs, a mixture of 80% methanol/20% water was used for the extractions reported here. The extracts were filtered and diluted 1:8 yielding test solutions containing 10% methanol, a level within the range that was found to be acceptable in the solvent-tolerance tests (Table 1). Yellow rice has typically been described as rice that has molded following dehulling and polishing. For this reason, the rice used for the recovery experiments was polished, dehulled, white rice. Because it was also expected that yellow rice samples would likely have been exposed to ambient lighting, and CTV is known to form an equilibrium with iso-CTV, a spiking solution was chosen that contained both CTV and significant iso-CTV (Figure 2B). Previous literature indicated that under ambient lighting CTV and iso-CTV reach an equilibrium with a ratio of approximately 1.5:1 to 2.3:1 CTV:iso-CTV. In the recovery studies reported here, the ratio of CTV:iso-CTV was 1.8:1. Rice was spiked with this mixture over the range of 0.36 to 7.24 mg/kg "total" (CTV + iso-CTV). Matrix matched calibration with the same stock solution was used to quantify the concentrations of total CTV + iso-CTV in the spiked rice.

While mAb 2-2 was the more sensitive of the two antibodies, in terms of IC_{50}s (Table 1, Figure S2), it also tended to give greater variability in response when toxin was absent (data not shown). For this reason, mAb 2-4 was selected as the basis of the screening assay for rice. Performance characteristics such as the limit of detection (LOD), IC_{20}, IC_{50}, and IC_{80} were determined from the calibration curve (Figure 3).

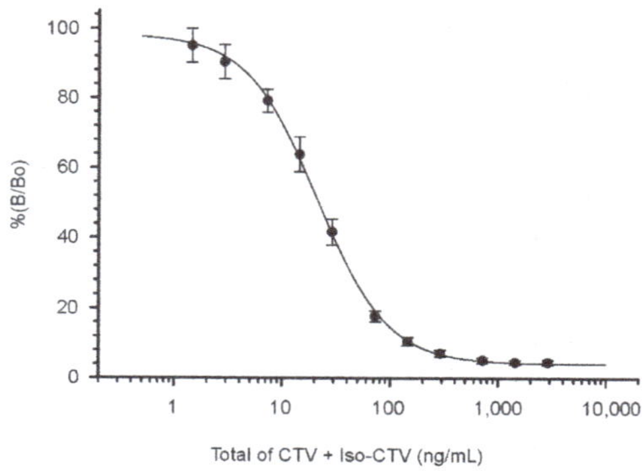

Figure 3. Calibration curve in rice matrix. The standard used for this curve contained a mixture of CTV and iso-CTV in the ratio of 1.8:1. Data are the mean values from 6 replicate plates with mAb 2-4. Error bars are ± 1 standard deviation (SD). The midpoint was 21.7 ng/mL, equivalent to 694 µg/kg in rice.

The LOD, calculated as the response three standard deviations below the response of the toxin-free samples, was estimated to be 0.13 mg/kg. The IC_{20} and IC_{80} were used to establish the dynamic range of 0.23 to 2.22 mg/kg. To improve quantification, spiked samples that contained greater than 2 mg/kg were additionally diluted to keep the resulting signals within the dynamic range. The recoveries observed were excellent (Table 3). From 30 samples covering the range of five spiking levels, the recoveries averaged 96.7% with individual recoveries that ranged from 74.0% to 118.8%. The relative standard deviation was 10.2%. We were not able to obtain any naturally contaminated samples of yellow rice that would permit evaluation beyond our recovery studies. CTV has been found at levels of up to 254 µg/kg in rice bran [4] which is at the lower end of the dynamic range, and in maize kernels at up to 2790 µg/kg [16], which is above the dynamic range of the mAb 2-4 CI-ELISA. Currently there are no regulatory levels established for CTV in commodities or foods, which makes determining a target level at which the assay should perform difficult. However, the data from oral toxicity testing provide a context for what levels may be important in foods. Recently a neurogenesis model was described wherein maternal mice were exposed to CTV through the diet from gestation day 6 through postnatal day 21. Male offspring were analyzed for effects on hippocampal neurogenesis. The no-observed-adverse-effect level was determined to be 1 mg/kg [31]. This level fits within the dynamic range for the CI-ELISA (0.23 to 2.22 mg/kg). Because of this, it appears that the CI-ELISA would be applicable to the screening of CTV and iso-CTV at levels that are toxicologically relevant.

Previous immunoassays for CTV have also reported good recoveries from rice and corn. In those experiments, the extraction solution was either 5% methanol (for rice powder) [23], or 10% to 20% methanol (for corn) [24,25]. Unfortunately, in all three cases the authors reported the spiking levels in units of "µg/mL", rather than in units related to the mass of the food (i.e., µg/kg or mg/kg). The units of µg/mL suggested that perhaps these referred to the spiking levels in liquid extracts, rather than to spiking levels in the solid food, a point that is unclear in the manuscripts. Because of this uncertainty, it is unclear what the real spiking ranges were in the previous work. Despite this, results from all three articles suggest recoveries were good, with average recoveries of 90% [24] 84–90% [25] and 70.5–95% [23]. Given the greater IC_{50}s for the previously reported antibodies (Table 2), it is likely that, on a weight basis, the spiking levels were generally higher than reported here.

Table 3. Recovery of a mixture of CTV and iso-CTV from spiked rice.

CTV (mg/kg)	Iso-CTV (mg/kg)	Total (CTV + Iso-CTV)	Average Recovery (% ± 1 SD) [a]	Number of Replicates [b]
0.23	0.13	0.36	91.9 ± 14.3	6
0.46	0.26	0.72	89.6 ± 7.8	6
0.93	0.52	1.45	97.9 ± 3.2	6
1.86	1.03	2.89	106.2 ± 8.7	6
4.65	2.58	7.23	97.7 ± 7.1	6
	Average (all):		96.7 ± 10.2	30

[a] Mean recovery ± one standard deviation (SD). [b] Number of rice samples that were spiked.

2.3. Effects of mAbs on the Fluorescence of Citreoviridin (CTV)

Molecular oxygen is known to quench most fluorophores [32]. Local environments that exclude molecular oxygen, for example the binding of a fluorophore within a protein, can increase fluorescence by reducing the inhibition caused by molecular oxygen. The previous sections have established that the mAbs selectively bound CTV. Therefore, we examined whether the fluorescence of CTV was influenced by the presence of mAbs or BSA. The concentrations of CTV and protein chosen for these experiments (1.25 and 2.0 μM, respectively) were selected so as to minimize interference from the fluorescence from the proteins themselves. The results demonstrated that the CTV mAbs significantly enhanced the fluorescence of CTV in aqueous systems (Figure 4).

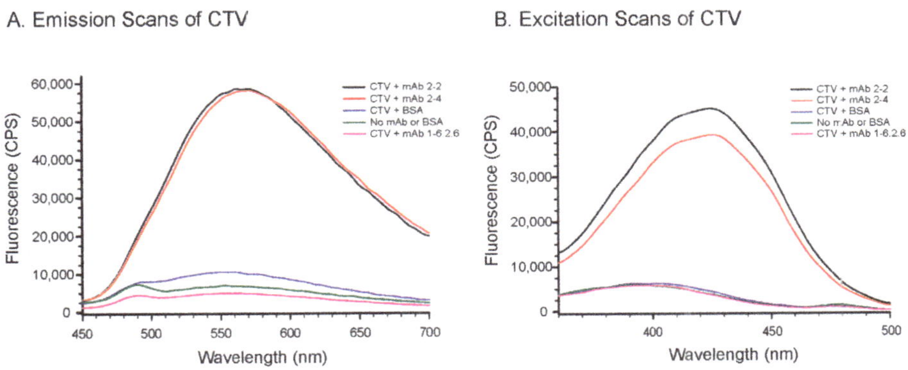

Figure 4. Effects of two CTV mAbs, bovine serum albumin (BSA), and a control mAb on the fluorescence of CTV. (**A**) Emission spectra with excitation at 420 nm. (**B**) Excitation spectra with emission at 570 nm. The concentration of CTV was 1.25 μM, while the concentrations of the proteins were 2.0 μM. The black and red lines represent CTV incubated with mAb 2-2 and mAb 2-4, respectively. The pink line represents the control of CTV incubated with an anti-deoxynivalenol mouse mAb (mAb 1-6.2.6). The green line represents the control of CTV in buffer without added mAb or BSA.

Both the excitation and emission of CTV were manifest as broad peaks. The excitation maximum was 420 nm and the emission maximum 570 nm (Figure 4). BSA also enhanced the fluorescence of CTV slightly. The concentrations of the proteins used in our experiments were relatively low. For immunoglobulin Gs 2.0 μM equates to approximately 0.3 g/L, while for BSA it equates to approximately 0.13 g/L. For comparison, the reference range for human serum albumin in blood is 35–50 g/L (approximately 525 to 750 μM), which is substantially higher. At those levels we suspect that the binding of CTV would also be increased significantly, although quantifying this effect would require separating out the fluorescence of the CTV from the previously reported fluorescence of the serum albumin [33]. With regard to the antibodies, the enhancement of CTV fluorescence was seen at levels at which the fluorescence of the antibody did not interfere (e.g., Figure 4). This observation could potentially be used to determine the strength of the binding interaction, something that we have only approximated here through competitive immunoassay. The effect might also be used to establish a fluorescence-based immunoassay for CTV, analogous to systems that have been reported for ochratoxin A and zearalenone [34].

3. Conclusions

Two mAbs that recognize CTV were developed and applied in CI-ELISAs to the detection of CTV and iso-CTV. With CTV in buffer, the assays were of good sensitivity, with IC_{50}s of 11 ng/mL (mAb 2-2) and 18 ng/mL (mAb 2-4). Assays based upon both mAbs were relatively tolerant to methanol and acetonitrile. For best results, it is recommended to keep the methanol concentration at or below 20% and the acetonitrile concentration at or below 10%. One of the mAbs (2-4) was applied to the detection of CTV and iso-CTV in spiked rice. Using matrix-matched calibration and a mixed CTV/iso-CTV standard, the dynamic range was equivalent to 0.23 to 2.22 mg/kg in rice. Recoveries were excellent, averaging 97 ± 10% over the range of 0.36 to 7.23 mg/kg. The developed mAbs and CI-ELISAs will be useful for the screening of CTV and iso-CTV in white rice. In addition, it was observed that both mAbs significantly enhanced the fluorescence of CTV, a phenomenon that may be useful in future efforts to determine the affinity of the antibodies for the toxin.

4. Materials and Methods

4.1. Materials

Except where noted otherwise, deionized water (Nanopure II, Thermo Scientific, Waltham, MA, USA) was used in the preparation of all reagents. The primary analytical standard of CTV was produced by Hayashi Pure Chemical Ind., Ltd. (Osaka, Japan). For spiking of rice samples, a standard material containing both CTV and iso-CTV prepared at the USDA-NCAUR (Peoria, IL, USA) was used [21]. Acetonitrile and methanol were HPLC grade and were purchased from Fisher Scientific (Hampton, NH, USA), as was polyvinyl alcohol (PVA). 1-1′carbonyldiimidazole (CDI), 4,6-dimethyl-2-oxo-2H-pyran-5-carboxylic acid (IDHA), 4,6-dimethyl-α-pyrone (DMP), and threose were purchased from Sigma-Aldrich (St. Louis, MO, USA). All other chemicals were reagent grade or better and purchased from major suppliers.

4.2. Liquid Chromatography-Photodiode Array-Mass Spectrometry (LC-PDA-MS) of CTV Stock Solutions

Two concentrated stock solutions containing predominantly CTV were prepared and analyzed by liquid chromatography-photodiode array-mass spectrometry (LC-PDA-MS) to determine relative purity. The first of the materials was prepared commercially and was used as the primary analytical standard. The second was prepared from material isolated by the late Robert Stubblefield at USDA-NCAUR (Peoria, IL, USA) and had been stored for many years at −20 °C. The latter was used to study the potential cross-reactivity of iso-CTV and was used to spike rice samples. A stock solution of the analytical standard was prepared in acetonitrile and a portion was diluted in methanol to a concentration of approximately 10 μg/mL. The ultraviolet–visible (UV–Vis) absorption spectrum was obtained using an Evolution 201 UV-Visible spectrometer (Thermo Scientific, Waltham, MA, USA). The absorbance at 383 nm was used to determine the concentration of total CTV and iso-CTV based upon the molar extinction coefficient of 44,000 [20]. A stock solution of the material used for spiking rice was prepared at nominally 0.6 mg/mL in acetone. The concentrations of CTV and iso-CTV in the spiking material were determined following LC-PDA-MS and were based upon the aforementioned analytical standard.

The LC instrument was a Dionex Ultimate 3000 system (ThermoFisher, Waltham, MA, USA). The column was a Kinetex 1.7 μ XB-C18, 100 Å, 100 × 3.0 mm, with a C18 Security Guard Ultra column (Phenomenex, Torrance, CA, USA) maintained at 30 °C. The mobile phase consisted of (A) acetonitrile, and (B) a mixture of 8% acetonitrile in water containing 0.25% acetic acid, with pH adjusted to 4 with ammonium hydroxide. The gradient began as 80% B for 4 min, then 65% B for 9 min. At 13 min the initial condition of 80% B was reinstated and the column was allowed to equilibrate for 7 min before injecting the next sample. For injection, the standards were diluted to 2 μg/mL. Injection volume was 20 μL. Flow rate was 0.7 mL/min, of which approximately 0.4 mL/min was directed to the PDA detector and 0.3 mL/min to the MS. The MS was a model QDa single quadrupole instrument (Waters Corp., Milford, MA, USA) operated in positive ion mode under the following parameters. Single ion monitoring at m/z = 403.2, cone voltage 6.0 V, probe temperature 425 °C, detector gain 1, capillary voltage 1.5 kV, sampling frequency 8 Hz. Under these conditions, CTV eluted at 7.8 min. At 8.5 min a second peak, with the same absorption spectrum as CTV eluted. In previous literature this peak has been termed iso-CTV, a convention also used here. Because no standards for iso-CTV were available the concentration of iso-CTV was calculated relative to that of the CTV analytical standard under the assumption that the ionization efficiencies for the two isomers were the same.

4.3. Preparation of CTV–Protein Conjugates

Purified CTV was linked to two proteins. One was used as a soluble antigen for immunizing mice and the other was used as an immobilized antigen in indirect ELISAs. The immobilized antigen, CTV-OVA2, was prepared as follows. All reactions were performed under conditions of reduced ambient lighting. CTV (5 mg) was dissolved in 0.25 mL acetone and 50 mg CDI was added and held at ambient temperature for 1 h. Twenty μL of water was added followed by 3.4 mL of OVA solution (50 mg OVA in 0.1 M $NaHCO_3$ buffer, pH 8.6). The mixture was shielded from light and stirred for 28 h at 4 °C. The product, CTV-OVA2, was dialyzed extensively against $NaHCO_3$ and phosphate buffers, using a 14 kDa membrane and culminating with 0.1 M PBS. The conjugate was diluted to a concentration of 2 mg/mL, freeze-dried, and stored at −80 °C until use. The immunogen, CTV-BSA3, was prepared in a similar fashion. However, instead of OVA the protein was BSA with a short ethylene diamine (EDA) linker attached. The preparation of EDA-BSA has been reported elsewhere [35].

4.4. Immunization of Animals and Isolation of mAb-Producing Clones

Immunization of mice, splenocyte fusions, cloning operations and antibody production were conducted at Envigo (Madison, WI, USA). Screening of sera and of fusion products were conducted at USDA-NCAUR (Peoria, IL, USA). All animal procedures were approved by the Institutional Animal Care and Use Committee (IACUC) of Envigo. Work was performed according to protocols 421-09a – Polyclonal Antibody Production and Hybridoma Development (approved 14 December 2018) and 638-18 Monoclonal Antibody Production (approved 14 December 2018). These protocols were developed in accordance with guidelines established by the U.S. National Institutes of Health- Office of Laboratory Animal Welfare. Ten female Balb/C mice were given a primary immunization of 100 μg CTV-BSA3 using the same procedures as described previously for production of paxilline antibodies [36]. Mouse antisera and hybridoma culture supernatant solutions were tested by competitive indirect ELISA (CI-ELISA). In initial experiments 0.1 mL of CTV-OVA2 (10 μg/mL in 0.05 M sodium phosphate buffer, pH 7.2), was incubated in wells of polystyrene microtiter plates overnight at 4 °C. The plates were washed twice with 0.32 mL Tween-PBS (0.02% *v*/*v* Tween-20 in 0.01 M PBS, pH 7.2) and blocked with 0.32 mL of PVA-PBS (1% *w*/*v* PVA in 0.01 M PBS) for 2 h at ambient temperature. Test solutions were prepared by mixing equal volumes of toxin standard solution (or control solution) and antiserum (alternatively, culture supernatant) diluted in BSA-PBS (1% *w*/*v* BSA in 0.01 M PBS) in the wells of a polypropylene microtiter plate (Corning Inc., Corning, NY, USA). The test plate containing immobilized antigen was washed twice with Tween-PBS and 0.1 mL of test solution was transferred to each well and incubated for 30 min at ambient temperature. The plate was then washed three times and 0.1 mL of a 1:2000 dilution of goat anti-mouse peroxidase conjugate was added. After incubating at ambient temperature for 30 min the plate was washed four times and the substrate o-phenylenediamine (OPD) was added. Preparation of the substrate is described elsewhere [36]. After 5 min 0.1 mL of 1 N hydrochloric acid was added to stop the reaction. Color development was determined by measuring the absorbance at 490 nm using a Synergy Neo microplate reader (Bio-Tek, Winooski, VT, USA).

From the 10 mice two were selected to undergo splenocyte fusion. The animals were sacrificed and the spleens aseptically removed. The splenocytes were chemically fused with NS-1 myeloma cells using polyethylene glycol then plated in HAT selection media. After 10 days, cultures were isolated and screened for anti-CTV activity using the CI-ELISA described earlier in this section. The first fusion yielded only a single product, which did not recognize free CTV. The second fusion also yielded a low number of products (5) of which three recognized free CTV. From these three products two clones were subsequently isolated, expanded, and used to produce larger amounts of antibody for evaluation. The cell lines were designated 2-2.2.2.8 (herein referred to as "2-2") and 2-4.3.5.1.2.1.4 (herein referred to as "2-4"). Ascites fluid from mice administered these cell lines was partially purified by ammonium sulfate precipitation using procedures described previously [37], then dialyzed against 0.1 M PBS and freeze-dried. The protein content was determined using the BCA Protein Assay (Thermo Fisher, Waltham, MA, USA).

4.5. Effects of Methanol and Acetonitrile on Two CTV mAbs

The impact of methanol and acetonitrile on the CI-ELISAs with mAbs 2-2 and 2-4 was determined by incorporating the solvents at concentrations from 5% to 30% (*v*/*v*) in the diluent used to prepare the calibration curves. The analytical CTV stock solution (i.e., 93% CTV, 7% iso-CTV) was used for these experiments. CI-ELISAs were performed essentially as in Section 4.4, with a lower concentration of CTV-OVA2 immobilized (4 μg/mL). Preliminary experiments indicated that using mAb 2-2 at a concentration of 12 μg/mL or mAb 2-4 at a concentration of 4.9 μg/mL were sufficient to give color development of approximately 1 absorbance unit with a 5 min substrate incubation. As described in Section 4.4 the antibodies (in 0.1% OVA-PBS) were mixed with the toxin/solvent combinations 1+1 before transferring them to the test plates containing immobilized CTV-OVA2. Note that at 30% methanol, the OVA was very poorly soluble and with 20% acetonitrile it was insoluble. For these reasons the highest solvent concentrations tested were 30% methanol and 15% acetonitrile. Tests were conducted with 4 to 8 replicate microtiter plates. Each CTV concentration level was represented by 4 wells on the plate. To obtain the most accurate comparisons, the two antibodies were compared side-by-side on the same plates. Standards were prepared over the range of 0.2 to 1000 ng/mL. Absorbance of all samples (i.e., "B") were normalized to those of the toxin-free controls (i.e., "Bo") using the equation (B/Bo) × 100%. Calibration curves based upon the transformed data were obtained using a logistic dose-response model and curve fitting software (TableCurve curve 2D, Systat Software Inc., Richmond, CA, USA). Fitted curves were used to calculate the concentrations required to inhibit color development by 50% (IC_{50}s).

4.6. Spiking and Extraction of Rice Samples

Polished long-grain white rice was kindly supplied by Susan P. McCormick (USDA-ARS-NCAUR). The rice had been milled so the husk, bran, and germ were removed. The rice was ground in a coffee mill to produce a powder similar in consistency to flour. Ten g of ground rice were spiked with a mixture of CTV and iso-CTV. The spiking solution contained a total of 144.7 μg/mL in methanol, of which 93 μg/mL was CTV and 51.7 μg/mL was iso-CTV. The volumes, amounts, and corresponding levels of CTV and iso-CTV in the rice are summarized in Table 4. Six replicate samples were produced at each spiking level. Samples were kept overnight at ambient temperature to permit the solvent in the spiking solution to evaporate. Samples were extracted with 40 mL of 80% (*v*/*v*) methanol/water by shaking for 2 h on a Burrell model 75 wrist-action shaker (Burrell Corporation, Pittsburgh, PA, USA) for 2 h at ambient temperature. The mixture was filtered through a Whatman 2V filter (Whatman plc, Maidstone, UK). The filtrate was diluted 1+7 (*v*/*v*) with a solution of 1% OVA in 10 mM PBS. The resulting diluted extracts contained 0.03125 g equivalents of rice/mL. Unspiked, control, rice was likewise extracted, diluted, and the diluted extract was used to prepare matrix-matched calibration curves for use in the CI-ELISA experiments.

Table 4. Preparation of spiked rice samples.

Volume of Stock Added (μL)	Amount of CTV Added (μg)	Amount of Iso-CTV Added (μg)	Concentration of CTV in Spiked Rice (mg/kg)	Concentration of Iso-CTV in Spiked Rice (mg/kg)	Concentration of CTV + Iso-CTV in Spiked Rice (mg/kg)
25	2.33	1.29	0.233	0.129	0.362
50	4.65	2.58	0.465	0.258	0.723
100	9.30	5.17	0.930	0.517	1.447
200	18.60	10.34	1.861	1.034	2.894
500	46.50	25.85	4.652	2.585	7.235

4.7. CI-ELISA of Rice Samples

Samples of rice spiked with a mixture of CTV and iso-CTV in the ratio of 1.8:1 were compared to calibration curves prepared using the same spiking mixture. The standards were prepared by diluting the stock solution in a control rice extract consisting of 0.03125 g equivalents of rice/mL (Section 4.6). As this was a 1:8 dilution of an 80% methanol extract of rice, the diluted extract also contained 10% methanol. The concentrations of the standards, represented as the sum of CTV and iso-CTV, ranged from 1.45 ng/mL to 2894 ng/mL. Because the matrix consisted of 0.03125 g equivalents/mL, the calibration curve covered the range from 0.046 to 93 mg/kg rice. To improve quantification, the spiked samples that contained greater than 2 mg/kg were diluted further to keep the resulting signals within the dynamic range of 80% to 20% of the maximal signals (i.e., between the IC_{20} and IC_{80}). The LOD of the assay in rice matrix was calculated by measuring the standard deviation of the unspiked (toxin-free) controls and calculating the CTV concentration required to observe a signal 3 standard deviations from the mean of the controls.

4.8. Effects of mAbs and Bovine Serum Albumin (BSA) on CTV Fluorescence

Preliminary experiments indicated that the mAbs could influence the fluorescence of CTV in aqueous buffer. To examine this further, mixtures of mAb or BSA were combined with CTV in aqueous buffer (10 mM PBS, pH 7.2). The final concentrations of mAb (or BSA) and CTV were 2.0 μM and 1.25 μM, respectively. The volumes of test solutions were 0.32 mL. Excitation and emission spectra were collected using a Neo microplate reader (BioTek, Winooski, VT, USA). Emission scans were collected using an excitation of 420 nm, with emission monitored over the range of 450 to 700 nm in 1 nm increments. Excitation scans were collected using an emission of 570 nm, with excitation provided over the range of 360 to 500 nm in 1 nm increments. Both excitation and emission data were collected with the top optics of the instrument and the following parameters: gain 150, normal reading speed, optics positioned 4.50 mm above the sample, and temperature 20.5 °C. Data from triplicate plates were averaged and imported into TableCurve. Spectra were smoothed using Savitzky–Golay filtering.

Supplementary Materials: The following are available online at http://www.mdpi.com/2072-6651/11/11/630/s1, Figure S1: Effect of iso-CTV on the calibration curve in diluted rice matrix using mAb 2-4, Figure S2: Effect of iso-CTV on the calibration curve in diluted rice matrix using mAb 2-2.

Author Contributions: C.M.M. and Y.S.-K. conceived and designed the experiments. C.M.M. conducted the experiments. Y.S.-K., N.K., Y.U., F.K., and C.M.M. contributed reagents/materials/analysis tools, and writing the manuscript.

Funding: This work was supported by USDA-Agricultural Research Service project number 5010-42000-0049-00D and by the Health and Labour Sciences Research Grants (Research on Food Safety, H28-shokuhin-ippan-004) from the Ministry of Health, Labour and Welfare of Japan.

Acknowledgments: The mention of trade names or commercial products in this publication is solely for the purpose of providing specific information and does not imply recommendation or endorsement by the U.S. Department of Agriculture. The USDA is an equal opportunity provider and employer.

Conflicts of Interest: The authors declare no conflict of interest. The founding sponsors had no role in the design of the study; in the collection, analyses, or interpretation of data; in the writing of the manuscript, and in the decision to publish the results.

References

1. Uraguchi, K. Mycotoxic origin of cardiac beriberi. *J. Stored Prod. Res.* **1969**, *5*, 227–236. [CrossRef]
2. Kushiro, M. Historical review of researches on yellow rice and mycotoxigenic fungi adherent to rice in Japan. *JSM Mycotoxins* **2015**, *65*, 19–23. [CrossRef]
3. Sakabe, N.; Goto, T.; Hirata, Y. The structure of citreoviridin, a toxic compound produced by *P. citreoviride* molded on rice. *Tetrahedron Lett.* **1964**, *5*, 1825–1830. [CrossRef]
4. Rosa, C.A.R.; Keller, K.M.; Oliveira, A.A.; Almeida, T.X.; Keller, L.A.M.; Marassi, A.C.; Kruger, C.D.; Deveza, M.V.; Monteiro, B.S.; Nunes, L.M.T.; et al. Production of citreoviridin by *Penicillium citreonigrum* strains associated with rice consumption and beriberi cases in the Maranhão State, Brazil. *Food Add. Contam. Part A* **2010**, *27*, 241–248. [CrossRef] [PubMed]
5. Lima, H.C.A.V.; Porto, E.A.S.; Marins, J.R.P.; Alves, R.M.; Machado, R.R.; Braga, K.N.L.; de Paiva, F.B.; Carmo, G.M.I.; e Santelli, A.C.F.S.; Sobel, J. Outbreak of beriberi in the state of Maranhão, Brazil: Revisiting the mycotoxin aetiologic hypothesis. *Trop. Dr.* **2010**, *40*, 95–97. [CrossRef] [PubMed]
6. Almeida, M.I.; Almeida, N.G.; Carvalho, K.L.; Gonçalves, G.A.A.; Silva, C.N.; Santos, E.A.; Garcia, J.C.; Vargas, E.A. Co-occurrence of aflatoxins B_1, B_2, G_1 and G_2, ochratoxin A, zearalenone, deoxynivalenol, and citreoviridin in rice in Brazil. *Food Add. Contam. Part A* **2012**, *29*, 694–703. [CrossRef]
7. Ueno, Y.; Ueno, I. Isolation and acute toxicity of citreoviridin, a neurotoxic mycotoxin of *Penicillium citreo-viride* Biourge. *Jpn. J. Exper. Med.* **1972**, *42*, 91–105.
8. Linnett, P.E.; Mitchell, A.D.; Osselton, M.D.; Mulheirn, L.J.; Beechey, R.B. Citreoviridin, a specific inhibitor of the mitochondiral adenosine triphosphatase. *Biochem. J.* **1978**, *170*, 503–510. [CrossRef]
9. Sayood, S.F.; Suh, H.; Wilcox, C.S.; Schuster, S.M. Effect of citreoviridin and isocitreoviridin on beef heart mitochondrial ATPase. *Arch. Biochem. Biophys.* **1989**, *270*, 714–721. [CrossRef]
10. Datta, S.C.; Ghosh, J.J. Production and purification of *Penicillium citreoviride* toxin and its effect on TPP-dependent liver transketolase. *Folia Microbiol.* **1981**, *26*, 408–412. [CrossRef]
11. Uchiyama, Y.; Takino, M.; Noguchi, M.; Shiratori, N.; Kobayashi, N.; Sugita-Konishi, Y. The in vivo and in vitro toxicokinetics of citreoviridin extracted from *Penicillium citreonigrum*. *Toxins* **2019**, *11*, 360. [CrossRef] [PubMed]
12. Ueno, Y. Temperature-dependent production of citreoviridin, a neurotoxin of *Penicillium citreo-viride* Biourge. *Jpn. J. Exper. Med.* **1972**, *42*, 107–114.
13. da Rocha, M.W.; Resck, I.S.; Caldas, E.D. Purification and full characterisation of citreoviridin produced by *Penicillium citreonigrum* in yeast extract sucrose (YES) medium. *Food Add. Contam. Part A* **2015**, *32*, 584–595. [CrossRef] [PubMed]
14. Wicklow, D.T.; Horn, B.W.; Cole, R.J. Cleistothecia of *Eupenicillium ochrosalmoneum* form naturally within corn kernels. *Can. J. Bot.* **1982**, *60*, 1050–1053. [CrossRef]
15. Peterson, S.W.; Jurjević, Ž.; Frisvad, J.C. Expanding the species and chemical diversity of *Penicillium* section Cinnamopurpurea. *PLoS ONE* **2015**, *10*, e0121987. [CrossRef]
16. Wicklow, D.T.; Stubblefield, R.D.; Horn, B.W.; Shotwell, O.L. Citreoviridin levels in *Eupenicillium ochrosalmoneum*-infested maize kernels at harvest. *Appl. Environ. Microbiol.* **1988**, *54*, 1096–1098.
17. Cole, R.J.; Dorner, J.W.; Cox, R.H.; Hill, R.A.; Cluter, H.G.; Wells, J.M. Isolation of citreoviridin from *Penicillium charlesii* cultures and molded pecan fragments. *Appl. Environ. Microbiol.* **1981**, *42*, 677–681.

18. Andrade, P.D.; Dantas, R.R.; Moura-Alves, T.L.d.S.d.; Caldas, E.D. Determination of multi-mycotoxins in cereals and of total fumonisins in maize products using isotope labeled internal standard and liquid chromatography/tandem mass spectrometry with positive ionization. *J. Chrom. A* **2017**, *1490*, 138–147. [CrossRef]
19. Tangni, E.K.; Pussemier, L. Ergosterol and mycotoxins in grain dusts from fourteen Belgian cereal storages: A preliminary screening survey. *J. Sci. Food Agric.* **2007**, *87*, 1263–1270. [CrossRef]
20. Cole, R.J.; Cox, R.H. *Handbook of Toxic Fungal Metabolites*; Academic Press: New York, NY, USA, 1981; p. 937.
21. Stubblefield, R.D.; Greer, J.I.; Shotwell, O.L. Liquid chromatographic method for determination of citreoviridin in corn and rice. *J. Assoc. Off. Anal. Chem.* **1988**, *71*, 721–724.
22. Engel, G. Untersuchungen zur bildung von mykotoxinen und deren quantitative analyse: VI. Citreoviridin. *J. Chrom. A* **1977**, *130*, 293–297. [CrossRef]
23. Zhuang, Z.H.; Que, S.J.; Gao, Y.M.; Yuan, J.; Ye, Z.; Du, M.; Lin, G.M.; Liu, L.C.; Wang, S.H. Artificial antigen synthesis and the development of polyclonal antibody-based immunoassay for citreoviridin determination. *World J. Microbiol. Biotechnol.* **2014**, *30*, 343–349. [CrossRef] [PubMed]
24. Jin, N.; Ling, S.; Yang, C.; Wang, S.H. Preparation and identification of monoclonal antibody against citreoviridin and development of detection by Ic-ELISA. *Toxicon* **2014**, *90*, 226–236. [CrossRef] [PubMed]
25. Wang, R.; Gu, X.; Zhuang, Z.; Zhong, Y.; Yang, H.; Wang, S. Screening and molecular evolution of a single chain variable fragment antibody (scFv) against citreoviridin toxin. *J. Agric. Food Chem.* **2016**, *64*, 7640–7648. [CrossRef]
26. Rebuffat, S.; Molho, D.; Dizabo, P. Mass spectra of citreoviridin A and related mycotoxins. *Org. Mass Spectrom.* **1984**, *19*, 349. [CrossRef]
27. Suh, H.; Wilcox, C.S. Chemistry of F1,F0-ATPase inhibitors. Stereoselective total syntheses of (+)-citreoviral and (−)-citreoviridin. *J. Am. Chem. Soc.* **1988**, *110*, 470–481. [CrossRef]
28. Nagel, D.W.; Steyn, P.S.; Ferreira, N.P. Biosynthesis of citreoviridin. *Phytochemistry* **1972**, *11*, 3215–3218. [CrossRef]
29. Hermanson, G.T.; Krishna Mallia, A.; Smith, P.K. *Immobilized Affinity Ligand Techniques*; Academic Press, Inc.: San Diego, CA, USA, 1992; p. 454.
30. ChemSpider. Available online: www.chemspider.com (accessed on 24 October 2019).
31. Nakajima, K.; Masubuchi, Y.; Ito, Y.; Inohana, M.; Takino, M.; Saegusa, Y.; Yoshida, T.; Sugita-Konishi, Y.; Shibutani, M. Developmental exposure of citreoviridin transiently affects hippocampal neurogenesis targeting multiple regulatory functions in mice. *Food Chem. Toxicol.* **2018**, *120*, 590–602. [CrossRef]
32. Lakowicz, J.R. *Principles of Fluorescence Spectroscopy*, 3th ed.; Springer: New York, NY, USA, 2006; p. 954.
33. Hou, H.; Qu, X.; Li, Y.; Kong, Y.; Jia, B.; Yao, X.; Jiang, B. Binding of citreoviridin to human serum albumin: Multispectroscopic and molecular docking. *Biomed. Res. Int.* **2015**. [CrossRef]
34. Li, T.; Jo, E.J.; Kim, M.G. A label-free fluorescence immunoassay system for the sensitive detection of the mycotoxin, ochratoxin A. *Chem. Commun.* **2012**, *48*, 2304–2306. [CrossRef]
35. Chu, F.S.; Lau, H.P.; Fan, T.S.; Zhang, G.S. Ethylenediamine modified bovine serum albumin as protein carrier in the production of antibody against mycotoxins. *J. Immun. Meth.* **1982**, *55*, 73–78. [CrossRef]
36. Maragos, C.M. Development and evaluation of monoclonal antibodies for paxilline. *Toxins* **2015**, *7*, 3903–3915. [CrossRef] [PubMed]
37. Maragos, C.M.; McCormick, S.P. Monoclonal antibodies for the mycotoxins deoxynivalenol and 3-acetyl-deoxynivalenol. *Food Agric. Immun.* **2000**, *12*, 181–192. [CrossRef]

toxins

MDPI

Article

A Rapid Extraction Method Combined with a Monoclonal Antibody-Based Immunoassay for the Detection of Amatoxins

Candace S. Bever [1], Robert M. Hnasko [2], Luisa W. Cheng [1,*] and Larry H. Stanker [1]

1 Foodborne Toxin Detection and Prevention Unit, Western Regional Research Center, Agricultural Research Service, United States Department of Agriculture, 800 Buchanan Street, Albany, CA 94710, USA; candace.bever@usda.gov (C.S.B.); larry.stanker@usda.gov (L.H.S.)

2 Produce Safety and Microbiology Unit, Western Regional Research Center, Agricultural Research Service, United States Department of Agriculture, 800 Buchanan Street, Albany, CA 94710, USA; robert.hnasko@usda.gov

* Correspondence: luisa.cheng@usda.gov; Tel.: +1-510-559-6337

Received: 22 November 2019; Accepted: 9 December 2019; Published: 11 December 2019

Abstract: Amatoxins (AMAs) are lethal toxins found in a variety of mushroom species. Detection methods are needed to determine the occurrence of AMAs in mushroom species suspected in mushroom poisonings. In this manuscript, we report the generation of novel monoclonal antibodies (mAbs, AMA9G3 and AMA9C12) and the development of a competitive, enzyme-linked immunosorbent assay (cELISA) that is sensitive at 1 ng mL^{-1} and shows selectivity for α-amanitin (α-AMA) and γ-amanitin (γ-AMA), and less for β-amanitin (β-AMA). In order to decrease the overall time needed for analysis, the extraction procedure for mushrooms was also simplified. A rapid (1 min) extraction procedure of AMAs using solvents as simple as water alone was successfully demonstrated using *Amanita* mushrooms. Together, the extraction method and the mAb-based ELISA represent a simple and rapid method that readily detects AMAs extracted from mushroom samples.

Keywords: amatoxins; amanitins; immunoassay; monoclonal antibodies; ELISA; death cap mushrooms

Key Contribution: Highly sensitive and selective monoclonal antibodies were developed to detect amatoxins. A rapid and simple extraction of amatoxins from mushroom specimens was demonstrated.

1. Introduction

There are thousands of reported mushroom poisonings occurring worldwide each year [1–5]. The most severe cases are from amatoxin (AMA)-containing mushrooms. AMA-containing mushrooms include a few species from the genera *Amanita*, *Galerina*, and *Lepiota*. The principle toxins responsible for the poisonings are the bicyclic octapeptides known as AMAs, most notably α-amanitin (α-AMA) and β-amanitin (β-AMA), and possibly γ-amanitin (γ-AMA) (Figure 1). In mice, the LD_{50} for α-AMA is 0.1 mg kg^{-1} [6] and, to humans, a dose of 0.3 mg kg^{-1} is severely toxic [7]. AMAs are potent inhibitors of RNA polymerase II, with bioactivity resistant to heat, cold, or acid inactivation. The typical distributions of α-AMA, β-AMA, and γ-AMA in a Death cap (*Amanita phalloides*) mushroom are approximately 43%, 43% and 14%, respectively [8,9]. A single dried mushroom typically contains around 1–2 mg g^{-1} of α-AMA [8,10,11].

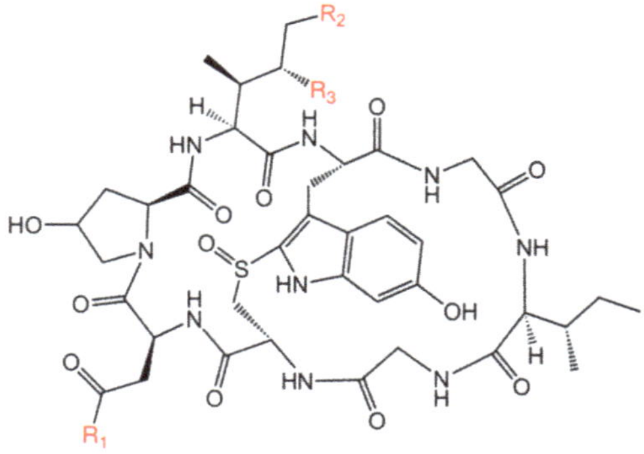

Compound	R_1	R_2	R_3
α-amanitin	NH_2	OH	OH
β-amanitin	OH	OH	OH
γ-amanitin	NH_2	OH	H

(a) **(b)**

Figure 1. Chemical structures of the amatoxin variants examined in this paper, (**a**) molecular structure of amanitin, (**b**) R-group designations for each variant.

The most common method for the detection of AMAs extracted from mushrooms is liquid chromatography (LC), coupled with UV detection or mass spectrometry (MS) [8,12–14]. Although these methods are sensitive and provide a high resolution of individual analytes, they are time-consuming and require expensive, laboratory-based instrumentation and highly trained personnel to interpret the results. In contrast, immunoassays are faster, can be field portable, and require less sophisticated instrumentation. The only commercially available antibody-based assay for AMA detection for research purposes is the Bühlmann assay [15]. This assay relies on a polyclonal antibody (pAb), which is a limited supply. Once the supply of antibody is depleted, the assay will have to be reevaluated for sensitivity and selectivity using a newly produced pAb. Since monoclonal antibodies (mAbs) are produced by a hybridoma cell line derived from a single cell, they overcome this supply limitation and have little or no batch-to-batch variability. Similarly, recombinant antibodies can be produced in large quantities, while preserving the monoclonality of the binding domain. Assays utilizing mAbs or recombinant antibodies are thus more desirable for long-term consistency and can be scaled-up for test kit manufacture. To our knowledge, only a few mAbs to AMAs have been described, and only one has been used for analytical detection [16–18].

Regardless of the method used to detect the toxin, extraction of the AMA is required before identification. Over the years, the extraction procedure has been streamlined from 24 h [8,10,19] to one hour [12,14,16,20]. Most of these methods have utilized an extraction solution consisting of methanol, acid, and water. Results from a latter study using a one hour extraction reported levels of α-AMA to be 0.88–1.33 mg g^{-1} dry weight [12], while earlier studies using the 24 hour extraction reported comparable levels of 0.75–2.8 mg g^{-1} dry weight [8,10] for the same species. Despite potential differences in the ages of mushrooms studied, these consistencies across studies suggest that extraction efficiency is not compromised with shortened extraction times. In addition, the historical methods use a combination of methanol, acid, and water to facilitate AMA extraction. Antibody-based immunoassays are often not compatible with large amounts of organic solvents or acidic solutions. Given the water solubility of AMAs, we hypothesized that a water-based AMA extraction would be sufficient for immunoassay detection.

The aim of this study was to utilize our previously reported immunogen, a periodate-oxidized form of α-AMA conjugated to the keyhole limpet hemocyanin (PERI-AMA-KLH) [20], to generate mouse mAbs. Then, we sought to use those mAbs to develop a sensitive and selective immunoassay for AMA detection from mushrooms. In this report, we describe and characterize novel anti-AMA mAbs and detail their performance in an indirect competitive inhibition enzyme-linked immunosorbent assay (cELISA). We compare the performance of this immunoassay for the detection of AMAs from mushrooms using difference extraction solutions. A sensitive detection assay for AMAs, combined with a rapid and simple toxin extraction method, would be a highly useful tool for the determination of AMA presence in wild mushrooms.

2. Results

2.1. Monoclonal Antibody Production

Mouse mAbs to AMAs were generated using the immunogen PERI-AMA-KLH [20]. Following the screening of the fusion plates, there were 14 positive cultures (optical density > 0.7), of which 12 cultures exhibited substantial signal reduction (optical density decreased by 0.5 or greater) in the presence of 100 ng mL^{-1} α-AMA in cELISA (Figure 2). Only two (9C12 and 9G3) of these grew stably, and were cloned multiple times until every well of the cell culture plate with cell growth elicited a positive indirect ELISA response to the coating antigen, a periodate-oxidized form of α-AMA conjugated to bovine serum albumin (PERI-AMA-BSA). The resulting mAbs were AMA9G3 (American Type Culture Collection Accession number PTA-125922) and AMA9C12 (American Type Culture Collection Accession number PTA-125923). Both mAbs were isotype IgG_1-possessing kappa light chains.

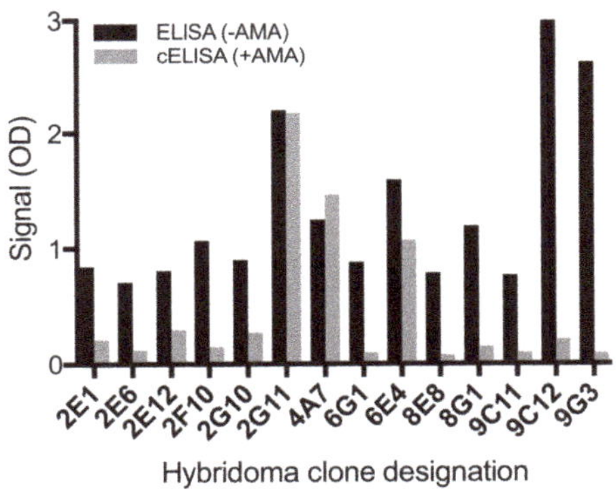

Figure 2. Hybridoma clone supernatants screened by indirect enzyme-linked immunosorbent assay (ELISA) (black bars) and by indirect competitive ELISA (gray bars). The cELISAs were completed using 100 ng mL^{-1} of α-amanitin as the competing analyte.

2.2. Cross-Reactivity and Sensitivity

In order to determine how effective the assay would be in selectively detecting AMAs, a panel of cyclic peptides and smaller chemicals was tested. These included the bicyclic heptapeptides known as phallotoxins (phalloidin and phallacidin) also produced by *A. phalloides*, chemical toxins (psilocybin, muscimol, and ibotenic acid) produced by other mushrooms, and cyclic peptides (nodularin and microcystin-LR) produced by cyanobacteria. Of these analytes tested, AMA9G3 was competitively inhibited by all of the AMAs, α-AMA, β-AMA, and γ-AMA, while AMA9C12 was only competitively inhibited by α-AMA and γ-AMA (Table 1 and Figure 3). Both mAbs did not bind to any of the other compounds tested.

Table 1. Cross-reactivity (%) of compounds found in associated mushrooms or structurally related compounds.

	mAb AMA9G3		mAb AMA9C12		Bühlmann Assay [15]
Toxin	**IC_{50} (ng mL^{-1})**	**Cross Reactivity (%)**	**IC_{50} (ng mL^{-1})**	**Cross Reactivity (%)**	**Cross Reactivity (%)**
α-amanitin	1.57 ± 0.07	100	2.66 ± 0.18	100	100
β-amanitin	24.2 ± 6.2	6.5	>1000	<0.3	0.1
γ-amanitin	1.63 ± 0.21	96	2.3 ± 0.31	115	90
phalloidin	>1000	<0.3	>1000	<0.3	Not detected
phallacidin	>1000	<0.3	>1000	<0.3	Not detected
psilocybin	>1000	<0.3	>1000	<0.3	Not determined
microcystin-LR	>1000	<0.3	>1000	<0.3	Not determined
nodularin	>1000	<0.3	>1000	<0.3	Not determined
ibotenic acid	>1000	<0.3	>1000	<0.3	Not determined
muscimol	>1000	<0.3	>1000	<0.3	Not determined

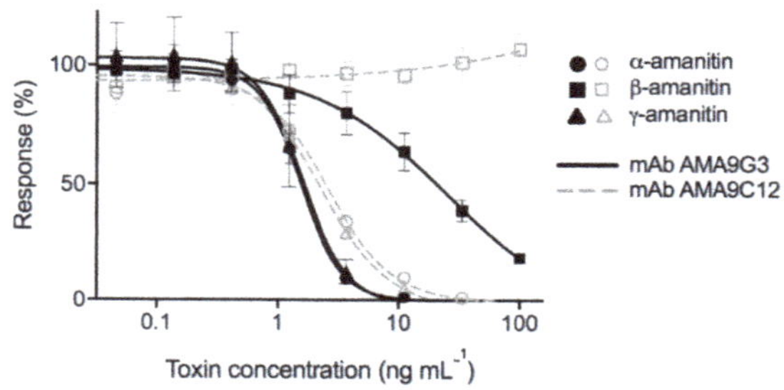

Figure 3. Standard cELISA inhibition curves for both monoclonal antibodies AMA9G3 (solid line) and AMA9C12 (dashed line) against toxins α-amanitin (circles), β-amanitin (squares), and γ-amanitin (triangles). $R^2 > 0.96$ for the linear portion (including a minimum of three points) for every curve.

The standard curves for the binding of both mAbs to the α-AMA, β-AMA, and γ-AMA toxins are shown in Figure 3. There is no reduction in signal response for AMA9C12 when tested against different concentrations of β-AMA, whereas, for AMA9G3, all three of the toxins did competitively inhibit at higher concentrations. The steepness of the curve indicated a small dynamic range for the assay because of the dramatic signal change produced by very small changes in toxin concentration, but simple sample dilutions could be performed to achieve a signal that fits within this range.

While both mAbs exhibited competitive inhibition from α-AMA and γ-AMA, AMA9G3 exhibited slightly higher sensitivity, with an IC_{50} of 1.57 ng mL^{-1} for α-AMA (Table 1 and Figure 3). For AMA9G3, the working range of detection (estimated as IC_{30} to IC_{80}) of α-AMA is 0.7–3.1 ng mL^{-1}, of γ-AMA is 0.5–2.4 ng mL^{-1}, and of β-AMA is 3.6–129.1 ng mL^{-1}. A conservative estimate for the limit of detection for α-AMA or γ-AMA with the AMA9G3 assay is 1 ng mL^{-1}, accounting for the large (30%) variation in signal at low to no concentrations of toxin. Because of the propensity for samples (mushroom extracts) to contain all three AMAs, AMA9G3 was selected for use in the cELISAs for the extraction studies.

2.3. Kinetic Measurements

For all three antibodies (two mAbs from this study and one rabbit pAb #58 from our previous work), a final concentration of 10 nM was used in both the Equilibrium and Kinetics Injection studies. Table 2 shows the affinity (K_d) and kinetic parameter (k_{on} and k_{off}) values obtained for each antibody tested against α-AMA as the free ligand. K_d values of 10^{-10} M and lower indicate that these antibodies exhibit a high affinity for their target analyte (α-AMA). Given the similar affinity characteristics between the mAbs and the pAb, the major advantage of the mAbs over the pAb is their ability to produce a continuous supply of the same protein.

Table 2. Affinities (K_d) and kinetic parameters (k_{on} and k_{off}) for antibodies binding to α-amanitin measured by KinExA.

Antibody	K_d (M)	k_{on} (M^{-1} s^{-1})	k_{off} (s^{-1})
rabbit pAb 58	3.5×10^{-11}	4.1×10^{6}	1.4×10^{-4}
mAb AMA9G3	6.4×10^{-11}	4.7×10^{7}	3.0×10^{-3}
mAb AMA9C12	9.3×10^{-10}	1.7×10^{7}	1.4×10^{-2}

2.4. Mushroom Extraction

For the purposes of exploring the feasibility of performing simplified and rapid extractions, five different extraction solutions were tested. The commonly employed extraction using methanol and dilute acid was compared to extractions with more innocuous reagents, such as phosphates, tris, and Tween-20. The extraction solutions were tested on three different mushroom species. Both *A. phalloides* and *A. ocreata* are known to contain AMAs, while *A. gemmata* is known to not contain AMAs.

AMAs were detected by this mAb-based cELISA in both of the extracts for the species known to contain AMAs (*A. phalloides* and *A. ocreata*) (Figure 4a,b). Detection is indicated by 100% inhibition at dilutions up to and including a 9000-fold dilution of the extract (Figure 4a,b). With increasing dilutions, it would be expected that the amount of inhibition would decrease as the AMA concentration decreases in a dose-dependent manner. Indeed, at dilutions greater than 9000-fold, decreased inhibition was observed for all extraction conditions.

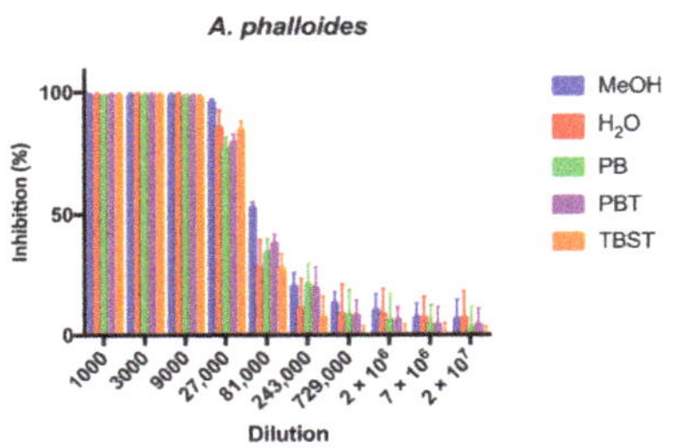

(a)

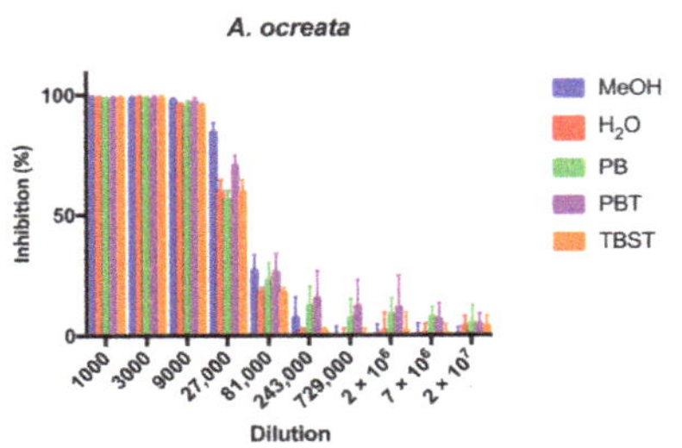

(b)

Figure 4. *Cont.*

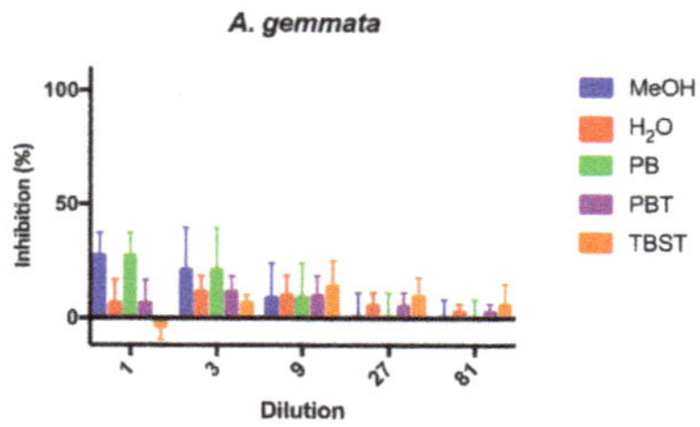

(**c**)

Figure 4. Representative cELISA inhibition profiles illustrating the amount of inhibition from the tested extraction conditions at varying dilutions of mushroom extracts obtained from (**a**) *Amanita phalloides*, (**b**) *A. ocreata*, and (**c**) *A. gemmata*. The extraction conditions were as follows: (1) MeOH: methanol: water: HCl for 1 h; (2) H_2O: diH_2O, 1 min, (3) PB: phosphate buffer, 1 min, (4) PBT: PB with Tween-20, 1 min, and (5) TBST: tris-buffered saline with Tween-20, 1 min.

There appeared to be very little difference between the different extraction conditions. The methanol extractions had a slightly higher inhibition at 27,000-fold and 81,000-fold dilutions for *A. phalloides* and the 27,000-fold dilution for *A. ocreata*. This may suggest that the methanol extractions extracted more AMAs than the aqueous extractions, since more inhibition equates to more toxin. However, the methanol extractions were carried out for 1 h, while the aqueous buffers were only carried out for 1 min. So, we cannot conclude whether time or composition contributed to the slight differences in extraction efficiency. However, because the differences were slight, we conclude that these aqueous buffers, with only a 1 min shaking step, were highly effective methods for AMA extraction.

With competitive-type assays, the signal intensity is the greatest when the free toxin to be detected is the lowest. In this case, low toxin levels are seen at the highest dilutions, and high signal intensities exhibit the most variation. With the data plotted as percent inhibition, higher variation was seen at higher dilutions (Figure 4). Thus, the background was estimated to be around 20–30% inhibition. Extracts from the non-toxin containing mushroom *A. gemmata* (Figure 4c) were intentionally tested in the more concentrated (e.g., 1- and 3-fold) dilutions, since no toxin was expected to be detected. The strong distinction between samples that exhibit 100% inhibition (i.e., with AMAs) and those without toxin demonstrated the ability of this mAb-based cELISA to selectively and sensitively detect AMAs from mushroom extracts.

3. Discussion

In this work, we report the generation of new anti-AMA mAbs (AMA9G3 and AMA9C12) using our previously synthesized immunogen (PERI-AMA-KLH) based on periodate oxidation of α-AMA. Unlike early reports of generating AMA-conjugated immunogens that exhibit toxicity [21,22], this immunogen did not cause any death in both mice (this study) or rabbits [20], corroborating the low toxicity observed by other investigators [23].

These mAbs exhibited high selectivity for AMAs, but the selectivity within AMAs varied from what was observed from our previously generated pAbs using the same immunogen [20]. The pAbs detected α-AMA, β-AMA and γ-AMA within a narrow range of IC_{50} values (2–3 ng mL^{-1}). Conversely, both mAbs distinguish the AMAs differently, such that AMA9G3 detected α-AMA and γ-AMA (IC_{50} ≈ 1.6 ng mL^{-1}), and, to a lesser extent, β-AMA (IC_{50} = 24 ng mL^{-1}), while AMA9C12 detected only α-AMA and γ-AMA (IC_{50} = 2.3 and 2.7 ng mL^{-1}, respectively). Similar to mAb AMA9C12, the commercially available, pAb-based cELISA detected only α-AMA and γ-AMA, and not β-AMA [15]. Another mAb, generated using a different coupling strategy to generate an α-AMA immunogen, produced mAbs with broad selectivity to α-AMA, β-AMA and γ-AMA (IC_{50} = 66, 97, and 163 ng mL^{-1}, respectively) [16]. A single chain variable fragment of a mAb, produced by this same group, also exhibited broad selectivity to α-AMA, β-AMA and γ-AMA (IC_{50} = 77, 115, and 199 ng mL^{-1}, respectively) [24].

Regarding total AMA extraction efficiency, the recovery obtained with all of these extraction methods is reputable. Previously reported α-AMA concentrations were approximately 1–2 mg of toxin per gram of dried mushroom [8,10,11], but this could vary depending on the location and age of the specimen [11]. Nonetheless, given our extraction ratio of 1 mL per 0.1 g of dried tissue, we would expect to recover 0.1–0.2 mg of α-AMA in 1 mL, as well as 0.1–0.2 mg of β-AMA and 0.05 mg of γ-AMA. This equates to extracts containing 10,000–45,000 ng mL^{-1} for total AMAs, which greatly exceeds our assay's detection limit of 1 ng mL^{-1} for total AMAs. Furthermore, the detection of AMAs in the 27,000-fold dilution samples of these AMA-containing sample extracts provided evidence that we were obtaining a high recovery of AMA. Other investigators performed a second extraction following a 1 h extraction (using an acetonitrile solution) and did not recover any detectable amounts of residual toxin in the second extract [14]. While most of the AMAs appear to be easily extracted, further work is needed to determine the recovery coefficient for the rapid extraction procedure.

Furthermore, none of these tissue samples were macerated prior to extraction for any of the solutions tested. Maceration of dried mushroom samples increases exposure of the researcher to the toxin-containing dust. In our previous work, and that of many others, the mushroom tissue was ground to a powder [11–14,19,20,25]. In this study, however, we did not grind the samples and still achieved sufficient toxin extraction, suitable for cELISA detection.

The sensitivity of our mAb-based cELISA permitted the detection of as little as 1 ng mL^{-1} of total AMAs in simple, water-based extracts obtained from dried mushrooms known to contain AMAs. Most LC methods achieve a detection limit of approximately 10 ng mL^{-1} for each individual AMA [8,14]. The amount of tissue required for an extraction can be as little as 50 mg for LC instrumental techniques, whereas the antibody-based detection methods could conceivably utilize sub-μg amounts of material using a simple, water-based extraction. The sensitivity, speed, and throughput of this mAb-based immunoassay for the detection of AMAs provides a simplified strategy for the evaluation of these toxins in wild mushrooms to characterize their occurrence and geographic distribution.

4. Materials and Methods

4.1. Immunization and Antibody Production

The Institutional Animal Care and Use Committee of the United States Department of Agriculture, Western Regional Research Center approved the experimental procedures used in these studies (protocol #16-1). Three 6-week-old female BALB/c mice were immunized by intraperitoneal injection (i.p.) of 100 μL of a 1:1 Sigma Adjuvant System (Sigma-Aldrich, St. Louis, MO, USA) containing 50 μg of PERI-AMA-KLH [20]. Two subsequent booster immunizations were administered i.p. at 2-week intervals using 20 μg of PERI-AMA-KLH in Sigma Adjuvant System. Serum were collected one week after the third immunization. Another two booster immunizations were performed four months later, two weeks apart, and serum was collected one week after this round of immunizations. After determining by indirect ELISA that the antibody response was still elevated to this target immunogen, a final booster immunization containing 10 μg of PERI-AMA-KLH in saline was administered i.p., four days prior to being euthanized and cell fusion.

4.2. ELISA Procedure

For serum antibody screening, black 96-well microtiter plates (Nunc, Thermo Fisher Scientific, Waltham, MA, USA) were coated at 1 μg mL^{-1} with PERI-AMA-BSA for 1 h at 37 °C in carbonate buffer (0.05 M carbonate-bicarbonate, pH 9.6). Then, the plates were blocked for 1 h at 37 °C with 3% non-fat dry milk in tris-buffered saline with 0.05% Tween-20 (TBST). After incubation for 1 h at 37 °C, TBST was removed and serum was loaded at a dilution of 1:100 in TBST and serially diluted. After another incubation for 1 h at 37 °C, plates were washed three times with TBST. Plates were then loaded with a secondary horse radish peroxidase labeled goat-anti-mouse antibody (Sigma) at 1:5000 in TBST and incubated for 1 h at 37 °C. Plates were washed and loaded with SuperSignal West Pico Chemiluminescent substrate (Fisher), incubated for 3 min, and then luminescent counts were recorded on a Victor^3 Multilabel Counter (PerkinElmer, Waltham, MA, USA).

4.3. Monoclonal Antibody Production and Screening

The cell fusion and expansion procedures were completed as previously described [26]. The screening of the cell culture plates following cell fusion, in particular the use of an indirect cELISA was carried out as previously described with minor modifications [27]. The screening process was the same, but the reagents used were changed. Briefly, wells of clear-bottom microtiter plates coated with PERI-AMA-BSA were pre-loaded with 50 µL/well of either TBST for noncompetitive screening or α-AMA at 100 ng mL^{-1} for competitive screening. Plates were incubated for 1 h at 37 °C, washed, and then incubated with a secondary antibody, as described for the direct screening. After incubation and washing, antibody activity was visualized using Enhanced K-Blue Substrate (Neogen, Lexington, KY, USA) and read on a VersaMax Microplate Reader (Molecular Devices, San Jose, CA, USA).

Hybridomas from wells exhibiting a significant reaction to the presence of α-AMA (i.e., a reduction in signal intensity) were selected for clonal expansion. Cells were cloned by limiting dilution, repeated until every well with cell growth presented positive activity via ELISA. MAbs were purified from the cell culture supernatant on a Protein G Sepharose affinity column (GE Healthcare Life Sciences, Pittsburgh, PA, USA), eluted with 0.1 M glycine-HCl, pH 2.7. Purified protein was extensively dialyzed against the phosphate buffered saline (PBS; 10 mM phosphate, 138 mM NaCl, 2.7 mM KCl, pH 7.4) and then stored at −20 °C until further use. Antibody protein concentrations were determined on a NanoDrop Lite Spectrophotometer (Thermo). Antibody isotyping was completed using an IsoStrip Mouse Monoclonal Antibody Isotyping Kit (Roche, Indianapolis, IN USA), following the manufacturer's protocol. Purified mAbs were titrated by indirect ELISA to determine the concentration of the antibody at half of the maximal signal. This determined concentration was used as the working concentration of the antibody for the cELISAs, to evaluate antibody cross-reactivity.

4.4. Antibody Characterization: Cross-Reactivity

Indirect cELISAs were completed using a panel of inhibitors to determine the selectivity of the mAbs. The cELISA procedure was nearly the same as that described for the serum screening, except for the addition of inhibitors (50 µL), which were mixed with 50 µL of the antibody solution during the primary antibody incubation step. The inhibitors tested were α-AMA (≥90%, Enzo Life Sciences, Farmingdale, NY, USA), β-AMA (≥90%, Enzo), γ-AMA (≥90%, Enzo), microcystin-LR (≥95%, Enzo), nodularin (≥95%, Enzo), phalloidin (>90%, Enzo), phallacidin (≥85%, Sigma), pysilocybin (>99%, Cerilliant, Round Rock, TX, USA), muscimol (>99%, Abcam, Cambridge, MA, USA), ibotenic acid (>98%, Abcam). Each analyte stock was dissolved in dH_2O, then serially diluted into TBST, starting at the highest concentration of ,000 ng mL^{-1}, and assessed in triplicate. Data were analyzed using a 4-parameter logistic equation (GraphPad Prism 7 Software, La Jolla, CA, USA) to determine the concentration of inhibition at half of the maximal signal (IC_{50}). Cross-reactivity (%) was calculated as follows: (IC_{50} α-AMA)/(IC_{50} test inhibitor) × 100.

4.5. Antibody Characterization: Kinetic Measurements

All KinExA experiments were performed on a KinExA 3200 with Autosampler (Sapidyne Instruments, Boise, ID, USA) and data were analyzed using KinExA Pro software provided by Sapidyne. Affinity values (K_d) utilized were their template protocol for an Equilibrium Experiment and kinetic parameters were determined using the Kinetics Injection method. Flow rates and volumes used were the default settings defined in the software.

Polymethylmethacrylate particles (aliquots of 200 mg, Syringa Labs, Boise, ID, USA) were adsorption coated with 30 µg of BSA-AMA-PERI in 1 mL of carbonate buffer for 1 h at room temperature with end-over-end rotation. The particles were blocked with a solution of 1% BSA (Sigma) in PBS for 1 h at room temperature with end-over-end rotation, and stored at 4 °C for no more than one week before use. The diluent for all reagents was PBS containing 1% BSA. Three antibodies were evaluated—two mouse mAbs (AMA9G3 and AMA9C12) generated from this study and one rabbit pAB #58 generated from the previous study [20]. The secondary antibody used for the mouse mAb experiments was DyLight650 labeled anti-mouse Ig (Fisher) (used at 0.5 μg mL^{-1}) and the secondary antibody used for the rabbit pAb experiments was AlexaFluor647 labeled anti-rabbit Ig (Jackson Immunoresearch, West Grove, PA, USA) (used at 0.25 μg mL^{-1}).

Signal test runs were completed on each antibody to determine the amount of antibody needed to generate the appropriate signal change (1 Δv). Then, for the Equilibrium experiments, antibody was prepared at 2× this concentration and then mixed with an equal volume of a solution containing α-AMA diluted 2-fold, ranging from 300 ng mL^{-1} (326 nM) to 9.2 pg mL^{-1} (10 pM) of final concentrations, including one sample with no α-AMA and one sample containing only diluent. For the Kinetics Injection experiments, the same 2× antibody concentration was used, along with solutions containing α-AMA diluted 2-fold, ranging from 920 ng mL^{-1} (1000 nM) to 1.8 ng mL^{-1} (2 nM). The Equilibrium and Kinetics Injection experiments were completed in duplicate.

4.6. Mushroom Extraction

Whole mushroom specimens were identified, dried, and provided by expert mycologists. The specimens included two that were known to contain AMAs, *A. phalloides* and *A. ocreata*, and one that was known to not contain AMAs, but was from the same genus, *A. gemmata*. Small portions of the specimens were weighed (~100–200 mg) and then placed into a 15 mL Falcon tube containing one of the five extraction buffers: (1) methanol (methanol:water:0.01 N HCl, 5:4:4, v:v:v), (2) diH_2O, (3) phosphate buffer (PB; 0.1 M, pH 7.6), (4) PB with Tween-20 (PBT), or (5) TBST at a ratio of 1 mL per 100 mg tissue. The samples that were extracted with the methanol buffer were shaken for 1 h at room temp and then centrifuged at 1000× *g* for 10 min. Aliquots of the supernatant were drawn off, diluted in TBST as necessary, and assessed by indirect cELISA. The samples in diH_2O, PB, PBT, or TBST were briefly shaken by hand for 1 min. Immediately after shaking, a 50 μL aliquot of the liquid phase was drawn off, diluted in TBST as necessary, and assessed by indirect cELISA, as described earlier. At least two individual mushrooms from each species were extracted, and extractions for each extraction condition were completed in duplicate.

Author Contributions: Conceived and designed the experiments, C.S.B. and L.H.S.; performed the experiments, C.S.B.; analyzed the data, C.S.B., L.W.C., R.M.H. and L.H.S.; wrote the original draft, C.S.B. and L.H.S.; reviewed and edited, L.W.C. and R.M.H.

Funding: This research was funded by the United States Department of Agriculture, Agricultural Research Service, National Program project NP108, CRIS 2030-42000-049-00D. C.S.B. and L.H.S. were also funded by interagency agreement IAA #60-2030-5-004 with Department of Homeland Security.

Acknowledgments: The authors are grateful to Tom Bruns and Catharine Adams (University of California, Berkeley) for their generous donation of identified mushrooms.

Conflicts of Interest: The authors declare a competing interest. C.S.B., R.M.H., L.W.C. and L.H.S. are named as inventors on a provisional patent application filed by ARS-USDA related to the data presented in this work.

References

1. Gummin, D.D.; Mowry, J.B.; Spyker, D.A.; Brooks, D.E.; Osterthaler, K.M.; Banner, W. 2017 Annual Report of the American Association of Poison Control Centers' National Poison Data System (npds): 35th Annual Report. *Clin. Toxicol.* **2018**, *56*, 1213–1415. [CrossRef] [PubMed]
2. Cervellin, G.; Comelli, I.; Rastelli, G.; Sanchis-Gomar, F.; Negri, F.; De Luca, C.; Lippi, G. Epidemiology and clinics of mushroom poisoning in northern Italy: A 21-year retrospective analysis. *Hum. Exp. Toxicol.* **2018**, *37*, 697–703. [CrossRef] [PubMed]

3. Jiang, L.; Luo, M.; Hao, L. Epidemiological characteristics of mushroom poisonings in Yunnan Province, China, 2004–2016. *Southeast Asian J. Trop. Med. Public Health* **2018**, *49*, 509–515.
4. Brandenburg, W.E.; Ward, K.J. Mushroom poisoning epidemiology in the United States. *Mycologia* **2018**, *110*, 637–641. [CrossRef] [PubMed]
5. Alagozlu, H.; Sezer, H.; Candan, F.; Tabak, E.; Elaldi, N. A survey of patients with acute poisoning in the Sivas region, Turkey, between 1994 and 1998. *Turk. J. Med. Sci.* **2002**, *32*, 39–42.
6. Wieland, T.; Wieland, O. Chemistry and toxicology of the toxins of *Amanita phalloides*. *Pharmacol. Rev.* **1959**, *11*, 87–107.
7. Yilmaz, I.; Ermis, F.; Kaya, E. A case study: What doses of *Amanita phalloides* and amatoxins are lethal to humans? *Wilderness Environ. Med.* **2015**, *26*, 491–496. [CrossRef]
8. Enjalbert, F.; Gallion, C.; Jehl, F.; Monteil, H.; Faulstich, H. Simultaneous assay for amatoxins and phallotoxins in *Amanita phalloides* fr. By high-performance liquid chromatography. *J. Chromatogr.* **1992**, *598*, 227–236. [CrossRef]
9. Vetter, J. Toxins of *Amanita phalloides*. *Toxicon* **1998**, *36*, 13–24. [CrossRef]
10. Kaya, E.; Karahan, S.; Bayram, R.; Yaykasli, K.O.; Colakoglu, S.; Saritas, A. Amatoxin and phallotoxin concentration in *Amanita phalloides* spores and tissues. *Toxicol. Ind. Health* **2015**, *31*, 1172–1177. [CrossRef]
11. Enjalbert, F.; Cassanas, G.; Salhi, S.L.; Guinchard, C.; Chaumont, J.P. Distribution of the amatoxins and phallotoxins in *Amanita phalloides*. Influence of the tissues and the collection site. *C. R. Acad. Sci. Ser. III* **1999**, *322*, 855–862. [CrossRef]
12. Sgambelluri, R.M.; Epis, S.; Sassera, D.; Luo, H.; Angelos, E.R.; Walton, J.D. Profiling of amatoxins and phallotoxins in the genus *Lepiota* by liquid chromatography combined with uv absorbance and mass spectrometry. *Toxins* **2014**, *6*, 2336–2347. [CrossRef] [PubMed]
13. Hu, J.; Zhang, P.; Zeng, J.; Chen, Z. Determination of amatoxins in different tissues and development stages of *Amanita exitialis*. *J. Sci. Food Agric.* **2012**, *92*, 2664–2667. [CrossRef] [PubMed]
14. Garcia, J.; Oliveira, A.; de Pinho, P.G.; Freitas, V.; Carvalho, A.; Baptista, P.; Pereira, E.; Bastos, M.D.L.; Carvalho, F. Determination of amatoxins and phallotoxins in *Amanita phalloides* mushrooms from northeastern Portugal by hplc-dad-ms. *Mycologia* **2015**, *107*, 679–687. [CrossRef]
15. Bühlmann Laboratories AG. *Amanitin ELISA Product Insert*; Catalog# ek-am1; Bühlmann Laboratories AG: Schönenbuch, Switzerland, 2013.
16. He, K.; Mao, Q.; Zang, X.; Zhang, Y.; Li, H.; Zhang, D. Production of a broad-specificity monoclonal antibody and application as a receptor to detection amatoxins in mushroom. *Biologicals* **2017**, *49*, 57–61. [CrossRef]
17. Faulstich, H.; Kirchner, K.; Derenzini, M. Strongly enhanced toxicity of the mushroom toxin alpha-amanitin by an amatoxin-specific fab or monoclonal antibody. *Toxicon* **1988**, *26*, 491–499. [CrossRef]
18. Baumann, K.; Munter, K.; Faulstich, H. Identification of structural features involved in binding of alpha-amanitin to a monoclonal antibody. *Biochemistry* **1993**, *32*, 4043–4050. [CrossRef]
19. McKnight, T.A.; McKnight, K.B.; Skeels, M.C. Amatoxin and phallotoxin concentration in *Amanita bisporigera* spores. *Mycologia* **2010**, *102*, 763–765. [CrossRef]
20. Bever, C.S.; Barnych, B.; Hnasko, R.; Cheng, L.W.; Stanker, L.H. A new conjugation method used for the development of an immunoassay for the detection of amanitin, a deadly mushroom toxin. *Toxins* **2018**, *10*, 265. [CrossRef]
21. Cessi, C.; Fiume, L. Increased toxicity of beta-amanitin when bound to a protein. *Toxicon* **1969**, *6*, 309–310. [CrossRef]
22. Faulstich, H.; Trischmann, H. Toxicity of and inhibition of rna polymerase by alpha-amanitin bound to macromolecules by an azo linkage. *Hoppe Seylers Z. Physiol. Chem.* **1973**, *354*, 1395–1398. [CrossRef] [PubMed]
23. Andres, R.Y.; Frei, W. [125I] Amatoxin and anti-amatoxin for radioimmunoassay prepared by a novel-approach: chemical and structural considerations. *Toxicon* **1987**, *25*, 915–922. [CrossRef]
24. He, K.; Zhang, X.Y.; Zhao, R.P.; Wang, L.X.; Feng, T.T.; Wei, D. An enzyme-linked immunosorbent assay and a gold-nanoparticle based immunochromatographic test for amatoxins using recombinant antibody. *Microchim. Acta* **2016**, *183*, 2211–2219. [CrossRef]
25. Stijve, T.; Seeger, R. Determination of α-, β- and γ-amanitin by high performance thin-layer chromatography in *Amanita phalloides* (vaill Ex fr.) secr, from Various Origin. *Z. Naturforsch.* **1979**, *34*, 1133–1138. [CrossRef] [PubMed]

26. Stanker, L.H.; Merrill, P.; Scotcher, M.C.; Cheng, L.W. Development and partial characterization of high-affinity monoclonal antibodies for botulinum toxin type A and their use in analysis of milk by sandwich elisa. *J. Immunol. Methods* **2008**, *336*, 1–8. [CrossRef]
27. Spier, C.R.; Bromage, E.S.; Harris, T.M.; Unger, M.A.; Kaattari, S.L. The development and evaluation of monoclonal antibodies for the detection of polycyclic aromatic hydrocarbons. *Anal. Biochem.* **2009**, *387*, 287–293. [CrossRef]

toxins

MDPI

Article

Rapid, Sensitive, and Accurate Point-of-Care Detection of Lethal Amatoxins in Urine

Candace S. Bever [1], Kenneth D. Swanson [2], Elizabeth I. Hamelin [2], Michael Filigenzi [3], Robert H. Poppenga [3], Jennifer Kaae [4], Luisa W. Cheng [1,*] and Larry H. Stanker [1]

1 Foodborne Toxin Detection and Prevention Research Unit, Western Regional Research Center, Agricultural Research Service, United States Department of Agriculture, 800 Buchanan Street, Albany, CA 94710, USA; candace.bever@usda.gov (C.S.B.); lstanker@gmail.com (L.H.S.)

2 Division of Laboratory Sciences, National Center for Environmental Health, Centers for Disease Control and Prevention, Atlanta, GA 30333, USA; ost3@cdc.gov (K.D.S.); ehamelin@cdc.gov (E.I.H.)

3 California Animal Health and Food Safety Laboratory System, University of California, 620 West Health Sciences Drive, Davis, CA 95616, USA; msfiligenzi@ucdavis.edu (M.F.); rhpoppenga@ucdavis.edu (R.H.P.)

4 Pet Emergency and Specialty Center of Marin, 901 E. Francisco Blvd, San Rafael, CA 94901, USA; jenkaae@gmail.com

* Correspondence: luisa.cheng@usda.gov; Tel.: +1-510-559-6337

Received: 30 January 2020; Accepted: 12 February 2020; Published: 15 February 2020

Abstract: Globally, mushroom poisonings cause about 100 human deaths each year, with thousands of people requiring medical assistance. Dogs are also susceptible to mushroom poisonings and require medical assistance. Cyclopeptides, and more specifically amanitins (or amatoxins, here) are the mushroom poison that causes the majority of these deaths. Current methods (predominantly chromatographic, as well as antibody-based) of detecting amatoxins are time-consuming and require expensive equipment. In this work, we demonstrate the utility of the lateral flow immunoassay (LFIA) for the rapid detection of amatoxins in urine samples. The LFIA detects as little as 10 ng/mL of α-amanitin (α-AMA) or γ-AMA, and 100 ng/mL of β-AMA in urine matrices. To demonstrate application of this LFIA for urine analysis, this study examined fortified human urine samples and urine collected from exposed dogs. Urine is sampled directly without the need for any pretreatment detection from urine is completed in 10 min, and the results are read by eye, without the need for specialized equipment. Analysis of both fortified human urine samples and urine samples collected from intoxicated dogs using the LFIA correlated well with liquid chromatography–mass spectrometry (LC-MS) methods.

Keywords: lateral flow immunoassay; amatoxins; amanitins; point-of-care; mushroom poisoning

Key Contribution: The study demonstrates the utility of the lateral flow immunoassay for the rapid detection of amatoxins in urine.

1. Introduction

Distinguishing toxic mushrooms from non-toxic ones is highly challenging, even for expert mycologists. Techniques to properly identify a mushroom include detailed morphological examination of the mushroom body, substrate identification, and knowledge of the location and the season The toxins often associated with lethal cases are cyclopeptides, and more specifically amanitins (most commonly α-amanitin (α-AMA), β-AMA, and γ-AMA, collectively referred to as amatoxins) [1] (Figure 1). Amatoxins are found in a few species of mushrooms from different genera, including *Amanita*, *Galerina*, and *Lepiota* [2]. Amatoxins are highly resistant to degradation, and on the cellular level they inhibit transcription by binding to RNA polymerase II. As little as 0.1 mg/kg body weight of amatoxins may cause death [3,4], and this amount can be found in a single *Amanita phalloides*.

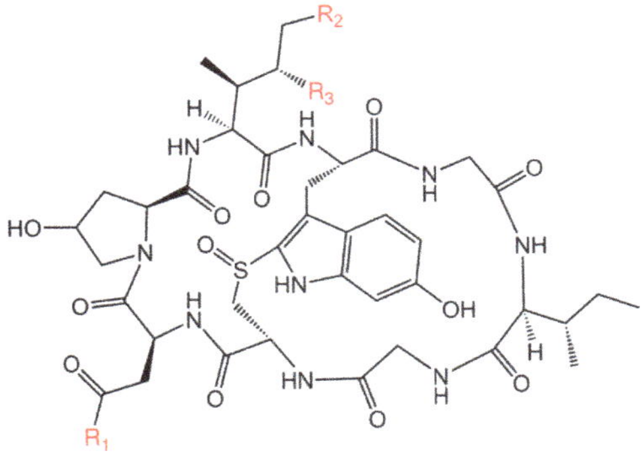

(a)

Compound	R_1	R_2	R_3
α-amanitin	NH_2	OH	OH
β-amanitin	OH	OH	OH
γ-amanitin	NH_2	OH	H

(b)

Figure 1. Chemical structures of the amatoxin variants examined in this paper, (**a**) molecular structure of the amanitin, (**b**) R-group designations for each variant.

Consumption of toxin-containing mushrooms can result in a range of symptoms, from mild to life-threatening [5,6]. The presumptive diagnosis for amatoxin poisoning is based on a history of consuming wild mushrooms (if known), presentation of delayed gastroenteritis, elevated liver enzyme levels, and ruling out other gastrointestinal diseases or conditions [6]. To distinguish amatoxin poisonings, the presence of an amatoxin in an intoxicated patient's urine would provide a definitive diagnosis. For dogs, obtaining a history of mushroom ingestion is rare, making diagnosis even more challenging. There are only a few laboratories capable of testing biological specimens for amatoxins to confirm human or animal exposures, and even when available, test results might not be available soon enough to help guide treatment. Although there are no FDA-approved antidotes, early diagnosis, aggressive immediate supportive care, and a range of potential therapies can potentially improve patient outcomes [6–11].

For both humans and dogs, the first symptoms of amatoxin poisonings usually appear 6–24 h after ingestion of an amatoxin-containing mushroom [6]. By this time, amatoxins have already begun damaging the liver and kidneys. Based on toxicological studies, amatoxins disappear rapidly from the serum, but are detectable in urine up to 4 days after ingestion [12–14]. In human urine, toxin concentrations decreased over time, and the highest concentrations observed were 4820 ng/mL for α-AMA and 7103 ng/mL for β-AMA [12]. Because of the relative ease of obtaining a urine sample, and the longer duration of detectability of amatoxins in urine compared to serum, urine seems an obvious sample matrix for performing rapid amatoxin analysis. Sensitive, rapid, and easy-to-perform methods are needed to detect amatoxins for the early diagnosis of toxin poisoning [9,15,16].

Current methods of chemical detection of amatoxins in urine include liquid chromatography–mass spectrometry (LC-MS) methods [17–23] and antibody-based enzyme-linked immunosorbent assays (ELISAs) [24,25]. LC-MS methods require sample extraction and expensive equipment, while ELISA methods require specialized equipment. Methods of both types typically take a few hours to complete. Lateral flow immunoassay (LFIA) formats utilize some of the reagents used for ELISA, but the entire test can be completed in minutes and requires no specialized equipment.

We have recently developed an LFIA for the detection of mushroom amatoxins [26]. As the sensitivity of the LFIA allows it to detect as little as 10 ng/mL, we hypothesized that this test would be useful for urine analysis in instances of mushroom poisonings. To test this hypothesis, we first conducted analysis of urine samples that were fortified with toxins (blind to the analyst) based on the reported concentrations of amatoxins identified in exposed individuals; and second, we used the LFIA to detect toxins in the urine samples collected from poisoned dogs. The LFIA results were compared to the established LC-MS methods [17,20]. Based on these studies, we can begin identifying the diagnostic utility of the LFIA for identifying amatoxin exposure.

2. Results and Discussion

2.1. Analytical Sensitivity of the LFIA and Interpretation of Results

Standard curves were obtained for the detection of α-AMA, β-AMA, and γ-AMA in a pooled urine matrix (Figure 2). As this is a competitive binding assay, the test line signal intensity decreases with the increase in toxin concentrations. The cut-off value for each individual amanitin was determined by selecting the concentration where the test line almost completely disappears, which is equivalent to a pixel intensity value of approximately 30. For α-AMA and γ-AMA, the cut-off value is 10 ng/mL, while for β-AMA, the cut-off is 100 ng/mL. Although difficult to discern by eye, the limit of detection (defined as three times the standard deviation of a sample without amatoxins) is 0.3 ng/mL for α-AMA and γ-AMA and 1 ng/mL for β-AMA. To ensure consistent interpretation of the line intensity by eye, the cut-off values determined in this study were used for the remainder of this study to determine the diagnostic accuracy of the test for urine analysis.

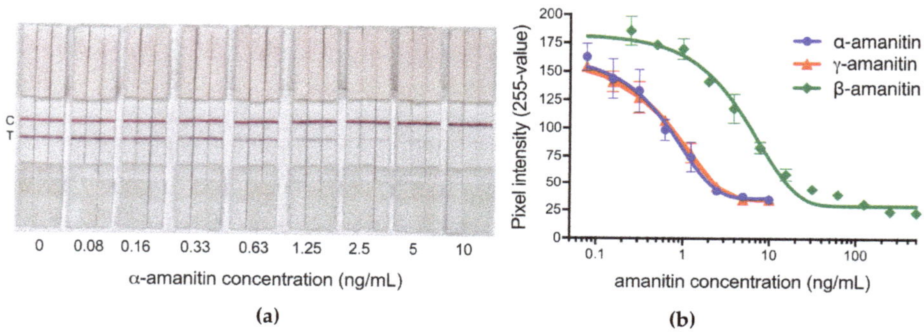

Figure 2. Standard curves for the detection of amatoxins by the lateral flow immunoassay (LFIA) in a pooled urine matrix. (**a**) A representative visual image of the LFIA test strips used for detecting α-amanitin (α-AMA). (**b**) Digitized values for the test line intensity for the detection of α-AMA, β-AMA, and γ-AMA. Data points represent the average of three replicates with error bars. T: test line, C: control line.

The cut-off value for β-AMA when detected in urine resulted in a 20-fold more sensitive assay cut-off value than our previous standard curve developed in phosphate-buffered saline (PBS) [26] There was no change in sensitivity for α-AMA or γ-AMA when using urine or PBS as the matrix [26] To test whether the β-AMA sensitivity differences were possibly due to the pH of the matrix, we evaluated pH-adjusted PBS buffers ranging from 4.5 to 8 (Figure 3). Indeed, there was a positive trend for the two concentrations of β-AMA tested (100 and 25 ng/mL), in which the test line intensity increased with the increase in pH. This trend corroborated with the differences between the pH of urine and that of PBS. The pH of the pooled urine matrix was 6.0, while the pH of PBS was 7.4.

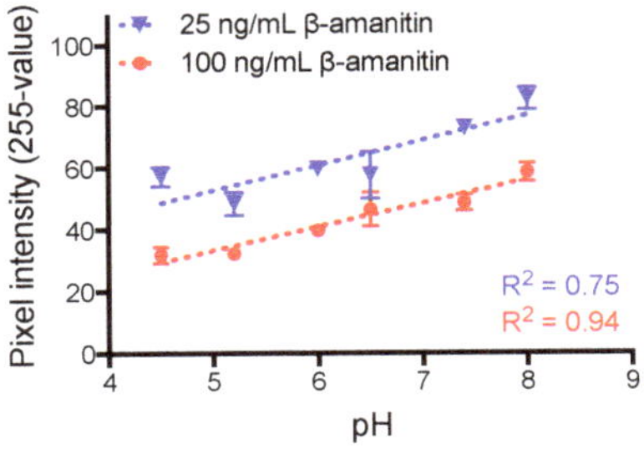

Figure 3. Test line intensities of the LFIA for solutions of β-amanitin in phosphate-buffered saline at different pHs.

2.2. Detection of Amatoxins in Blind Fortified Human Urine Samples

We recapitulated a human exposure study by spiking unexposed human urine samples with concentrations (45 to 4550 ng/mL) of amatoxins measured from actual food exposures [12]. Two modifications were made to the sample set: (1) we included γ-AMA along with α-AMA and β-AMA, and (2) we utilized a mixture of single and pooled urine samples. LFIA analysis was conducted as a blind test so that the LFIA readers would utilize only the LFIA as the detection method to see how well they could identify amatoxin-containing urine samples. All samples were also validated using an LC-MS/MS method [17] for confirmation. Figure 4 shows the nominal spiked concentrations along with LFIA and LC-MS/MS results.

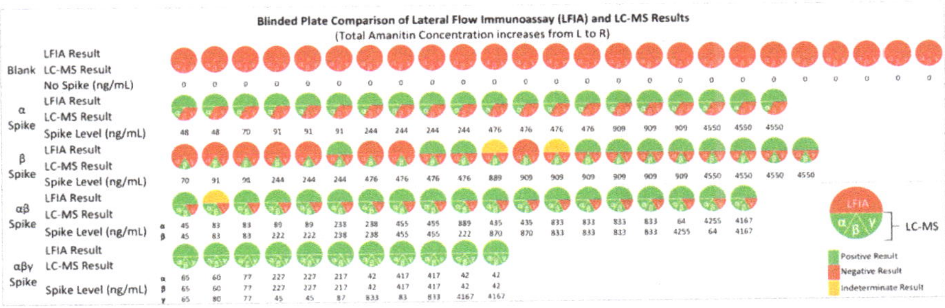

Figure 4. Design and results of experiments on amatoxin-fortified human urine samples. The toxin concentrations shown are the nominal spiked amounts. LC-MS: liquid chromatography–mass spectrometry, L: left, R: right.

In this sample set, the diagnostic efficiency of the LFIA for indicating when a sample contained amatoxins and when it did not was outstanding (94.6%) (Table 1). Diagnostic sensitivity (true positive rate) was 92.3% and specificity (true negative rate) was 100%. These diagnostic descriptor calculations took into account the cut-off values determined for each analyte (i.e., 10 ng/mL for α-AMA and γ-AMA, and 100 ng/mL for β-AMA). Overall, there were only three samples that were recorded as indeterminate due to the difference in interpretation by two independent readers.

Using the cut-off values for each analyte compared to the nominal spike value resulted in a designation of "false negative" for only five samples. A false negative result meant that the LFIA reported a negative result, although the nominal concentration was greater than the cut-off value that should have resulted in a positive detection result. All five false negative designations occurred in samples that contained only β-AMA in the amount of 244–909 ng/mL (Figure 4). Of these five incorrectly identified samples, three were from single urine samples, while the other two were from the pooled urine matrix. This error is plausibly due to the interpretation of a faint test line as the toxin concentration approaches the cut-off value of the assay.

Interestingly, 7 samples fortified with only β-AMA in the range of 244–909 ng/mL tested positive and were reported as true positive, since they matched the criteria of being above the assay's cut-off value. All the other samples fortified with only β-AMA below this range (i.e., 91 ng/mL or lower; n = 3) tested negative, and the samples that were above this range (i.e., 4550 ng/mL; n = 4) tested positive, as expected. Overall, inaccuracy of LFIA results is observed when only β-AMA is detected at a level close to its cut-off value. For α-AMA and γ-AMA, the concentrations tested were all at least 4-fold higher than the cut-off value, since these concentrations were those that were previously reported for amatoxin poisoning cases, and thus are clinically relevant. Detecting low concentrations of β-AMA may appear to be a limitation of this technology; however, α-AMA and β-AMA are almost always found at comparable concentrations in mushrooms [27,28], and thus either analyte serves as a good biomarker for determining amatoxin poisoning.

Table 1. Performance of LFIA for qualitative determination of the presence of amatoxins in blind fortified urine samples and collected dog urine samples.

Diagnostic Parameter	Fortified Human Urine [a]	Intoxicated Dog Urine [b]
# of samples	n = 96 [c]	n = 38
True positive (TP)	60	8
True negative (TN)	28	22
False positive (FP)	0	0
False negative (FN)	5	8
Sensitivity	92.3%	50%
Specificity	100%	100%
Efficiency	94.6%	78.9%

[a] Compared to the LC-MS/MS method [17] for confirmation. [b] Compared to LC-MS/MS/MS method [20] for confirmation. [c] 3 samples were not included in this analysis, because the LFIA results obtained by two independent readers were ambiguous.

Most previous amatoxin exposure studies measured α-AMA and sometimes β-AMA, however γ-AMA was often not included. Although the kinetics of γ-AMA are not well studied, we sought to include it because it could conceivably be a diagnostic marker of amatoxin poisoning. The previously described studies relying on immunoanalytical methods might also have been detecting γ-AMA because their reagents cross-reacted with this analyte [25]. Furthermore, the previously described immunoanalytical methods were less sensitive to β-AMA [29], thus those tests would have also missed the samples that this LFIA missed when only β-AMA was present in a urine sample.

The use of pooled and single urine samples was meant as a means to identify any potentially interfering components found in urine. In the instances of inaccuracy mentioned above, no differences were attributed to those samples being a single or a pooled urine sample. Furthermore, there were no false positives observed in this amatoxin-fortified human urine study, which means that when a sample had no amatoxins below the defined threshold concentration, it was interpreted correctly by the LFIA. Together, these results underscore that no apparent urine components (natural or potentially synthetic—urine samples were neither subjected to drug screening nor were they deemed free of drugs) interfere with amatoxin detectability.

2.3. Detection of Amatoxins in Dog Urine Samples

Unfortunately, many dogs each year are poisoned by amatoxins due to their natural curiosity and indiscriminate eating habits [30,31]. To evaluate this LFIA for diagnostic potential, we collaboratively analyzed samples submitted to the California Animal Health and Food Safety Laboratory in Davis, CA, USA (CAHFS Davis). Urine samples were qualitatively analyzed both by LFIA and LC-MS/MS/MS methods [20] (Table 1). These samples were collected from the dogs presumed to have ingested amatoxin-containing mushrooms and from the dogs not suspected of mushroom poisoning (a mix of healthy and sick dogs).

The LC-MS/MS/MS method only detects α-AMA. The LC-MS/MS/MS result reports "positive" if the sample has detectable (above 1 ng/mL) amounts of α-AMA, reports "negative" when no amounts are detected, and reports "trace" when a feature is detected with the correct retention time, molecular weight, and fragmentation pattern, but the concentration is below 1 ng/mL.

As shown in Figure 5, the LFIA should and does indicate negative when the LC-MS/MS/MS indicates trace ($n = 7$), since the limit of detection for the LC-MS/MS/MS is 1 ng/mL, which is 10-fold lower than for the LFIA. Only one sample was found to be negative by the LFIA and positive by the LC-MS/MS/MS. This particular sample had an estimated LC-MS/MS/MS concentration of 2 ng/mL, which is below the LFIA's cut-off limit (10 ng/mL), but above the threshold for positive for the LC-MS/MS/MS (1 ng/mL). However, for the rest of the samples that had either detectable amounts of α-AMA as determined by the LC-MS/MS/MS (i.e., true positive, $n = 8$) or non-detectable amounts of α-AMA as determined by the LC-MS/MS/MS (i.e., true negative, $n = 22$), the LFIA correlates 100% (Table 1). Based on this consistency, there seems to be no components in dog urine that interfere with generating reliable results. Based on this dog urine sample set, the calculated values for diagnostic sensitivity (50%), specificity (100%), and efficiency (78.9%) are provided in Table 1. This sample set distribution is not the representative distribution of prevalence of the poisonings encountered in the population, and so a larger sample size would help to determine diagnostic characteristics more accurately.

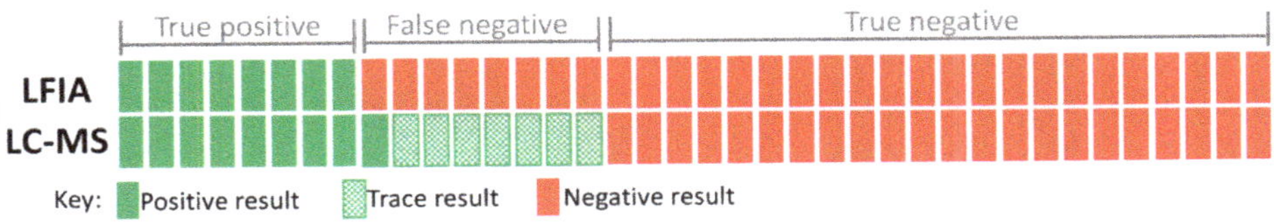

Figure 5. Comparison of methods (LC-MS/MS/MS and LFIA) for determining the presence of amatoxins in intoxicated dog urine samples ($n = 38$).

All detection methods have their benefits and limitations. LC-MS methods can provide a more definitive analysis for the samples containing lower concentrations of amatoxins, but all samples (e.g., urine and mushrooms) require sample extraction before detection. This LFIA method is exceptionally rapid (10 min), requires no sample extraction for urine, and the test is portable. In the mushroom poisoning scenarios where the illness progresses rapidly, the LFIA described here is distinctively appropriate for point-of-care urine testing. The speed of analysis and lack of requirement for trained personnel and expensive instrumentation make this an ideal point-of-care method. Because there is no clinical tool to determine amatoxin poisoning, this LFIA test should be further exploited given its reliable diagnostic performance in this study.

3. Materials and Methods

3.1. Materials and Approvals

All the unexposed human urine samples were obtained from a commercial provider (Tennessee Blood Services, Memphis, TN, USA) and pre-screened by the vendor in accordance with FDA regulations, and thus no consent procedures were required for this project (IRB #201210385). All animal urine samples were collected and submitted by the owners with their consent or by their veterinarians to the California Animal Health and Food Safety Laboratory System, Davis, and thus did not require the Institutional Animal Care and Use Committee review.

Monoclonal antibody (AMA9G3), hapten-protein conjugate (LB-AMA-BSA), and full LFIA test strips were produced as described earlier [26,32,33]. The LFIA components are diagramed in Figure 6.

The standards used were α-amanitin (α-AMA; ≥90%, Enzo Life Sciences, Farmingdale, NY, USA), β-AMA (≥90%, Enzo), and γ-AMA (≥90%, Enzo).

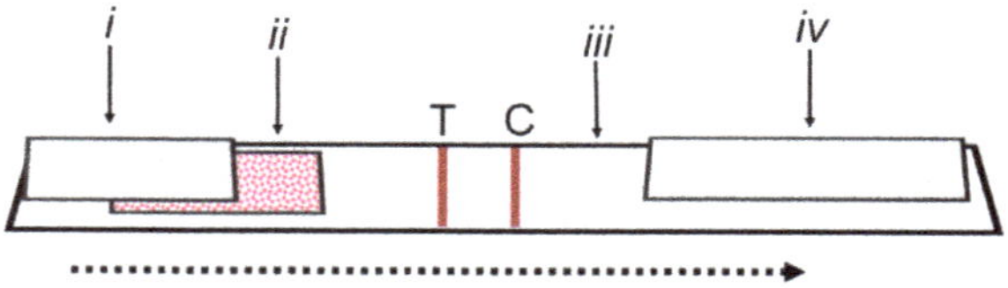

Figure 6. Schematic diagram of the lateral flow strip components: (i) sample pad, (ii) conjugate pad, (iii) nitrocellulose membrane, (iv) wicking pad, (T) test line, (C) control line. The arrow indicates the flow direction.

3.2. *Analytical Sensitivity of the LFIA*

The sensitivity of the LFIA for α-AMA, β-AMA, and γ-AMA was determined using a pooled urine matrix. The urine matrix for calibration curves was generated using 10 single urine samples pooled together. All 10 single urine samples were tested separately as negative controls and confirmed to not interfere with the read-out. Two-fold dilutions of the standards ranging from 0.08 to 2000 ng/mL were made using the pooled urine as the diluent. Then, 100 μL of each concentration were applied to the sample pad region of the test strip and tested in triplicate. The signal intensity of both the control and test lines was resolved in 10 min. If no visible control line appeared on the test strip, the test was determined to be invalid. Digital photographs of the test strips were obtained using a Nikon SLR camera equipped with an LED ring light (B&H Foto and Electronics Corps, New York, NY, USA) for even lighting. The images were analyzed using ImageJ software (NIH, Bethesda, MD, USA). Images were contrast-enhanced (default setting of 0.3%), and boxes of consistent size were used to integrate the test line's pixel value. Pixel values were inverted by subtracting the measured value from the maximum possible (i.e., 255). Data was plotted using GraphPad Prism 7 software (GraphPad Software, San Diego, CA, USA) using a four-parameter logistic equation.

PBS solutions (10 mM phosphate, 138 mM NaCl, 2.7 mM KCl, pH 7.4) were adjusted to pH 4.5, 5.2, 6, 6.5, 7.4, and 8 by adding either 1 M NaOH or 1 M HCl. Each solution was then spiked with β-AMA at 100 and 25 ng/mL. As described above, 100 μL of each solution were applied to the sample pad region of the test strip, tested in triplicate, incubated for 10 min, photographed, and interpreted by digital analysis.

3.3. *Amatoxin-Fortified Human Urine Analysis*

Single (*n* = 48) and pooled (*n* = 48) human urine samples (obtained under IRB #201210385) were fortified (blind to the LFIA researchers) with varying amounts of α-AMA, β-AMA, and γ-AMA. A 96-well microtiter plate was prepared by spiking urine with α-AMA, β-AMA, and γ-AMA at concentrations ranging from 45 to 4550 ng/mL following earlier findings from exposed individuals [12]. The distribution of samples was intended to follow these findings such that 19 wells contained α-AMA and β-AMA, 20 wells contained only α-AMA, 21 wells contained only β-AMA, 11 wells contained α-AMA, β-AMA, and γ-AMA, and 25 wells contained no amatoxins. These spiked concentrations were randomly distributed between the single and pooled urine matrices.

For analytical confirmation, all human urine samples were also analyzed following the previously validated LC-MS/MS method used for human urine [17]. The LFIA method was performed by placing 100 μL of a urine sample directly onto the sample pad and waiting for approximately 10 min before interpreting the line intensity of the test line. A visual qualitative reading of either YES(+) or NO(−) (YES – no visible test line, NO – a visible test line) was performed by two individuals, and a digital image of the strip was also acquired. If no control line appeared, then the test was determined to be invalid. The LFIA's cut-off values were set at 10 ng/mL for α-AMA and γ-AMA, and at 100 ng/mL for β-AMA. If amatoxin concentrations exceeded these values, the result was to read positive (defined as true positive), and if the amatoxin concentrations were below these cut-off values, the result was to read negative (defined as true negative). On the basis of these studies, diagnostic sensitivity (true positive rate), diagnostic specificity (true negative rate), and diagnostic efficiency of the LFIA results were calculated using the following formulae:

False positive (FP) =	LFIA positive, although the amatoxin concentration was below the cut-off value
False negative (FN) =	LFIA negative, although the amatoxin concentration was above the cut-off value
Diagnostic sensitivity =	TP/(TP + FN)
Diagnostic specificity =	TN/(TN+ FP)
Single-study diagnostic efficiency =	(TP + TN)/(TP

3.4. *Collected Dog Urine Analysis for Amatoxins*

Urine samples of the dogs with suspected amatoxin intoxication were submitted by their owners or by their veterinarians. For analytical confirmation, dog urine samples were analyzed following the previously validated LC-MS/MS/MS method used for dog urine [20]. The LFIA method was performed as described previously for human urine samples by placing 100 μL of a urine sample directly onto the sample pad and waiting for approximately 10 min before interpreting the line intensity of the test line.

Author Contributions: C.S.B., K.D.S., E.I.H., and L.H.S. designed research; C.S.B., M.F., K.D.S., and E.I.H. performed research; R.H.P., M.F., K.D.S., E.I.H., and J.K. contributed samples and analytic tools; C.S.B., K.D.S., E.I.H., M.F., L.W.C., and L.H.S. analyzed data; C.S.B. wrote the original draft; J.K., R.H.P., M.F., K.D.S., E.I.H., L.W.C., and L.H.S. reviewed and edited the manuscript. All authors have read and agreed to the published version of the manuscript.

Funding: This research was funded by the United States Department of Agriculture, Agricultural Research Service, National Program project NP108, CRIS 2030-42000-049-00D, and the USDA's Innovation Fund Award.

Acknowledgments: We thank Gary Richter and Mike Barlia for supplying dog urine specimens. Many thanks to Thomas Henderson, Paula Do, and Paul Merrill for their technical assistance.

Conflicts of Interest: The authors declare no conflict of interest.

Disclaimer: The findings and conclusions in this study are those of the authors and do not necessarily represent the views of the U.S. Department of Health and Human Services, or the U.S. Centers for Disease Control and Prevention. Use of trade names and commercial sources is for identification only and does not constitute endorsement by the U.S. Department of Health and Human Services, or the U.S. Centers for Disease Control and Prevention.

References

1. Brandenburg, W.E.; Ward, K.J. Mushroom poisoning epidemiology in the United States. *Mycologia* **2018**, *110*, 637–641. [CrossRef] [PubMed]
2. Walton, J.D. *The Cyclic Peptide Toxins of Amanita and Other Poisonous Mushrooms*, 1st ed.; Springer Science & Business Media: New York, NY, USA, 2018.
3. Kaya, E.; Karahan, S.; Bayram, R.; Yaykasli, K.O.; Colakoglu, S.; Saritas, A. Amatoxin and phallotoxin concentration in *Amanita phalloides* spores and tissues. *Toxicol. Ind. Health* **2015**, *31*, 1172–1177. [CrossRef] [PubMed]
4. Yilmaz, I.; Ermis, F.; Akata, I.; Kaya, E. A case study: What doses of *Amanita phalloides* and amatoxins are lethal to humans? *Wilderness Environ. Med.* **2015**, *26*, 491–496. [CrossRef]
5. Graeme, K.A. Mycetism: A review of the recent literature. *J. Med. Toxicol.* **2014**, *10*, 173–189. [CrossRef] [PubMed]
6. Benjamin, D.R. *Mushrooms: Poisons and Panaceas: A Handbook for Naturalists, Mycologists, and Physicians*; W. H. Freeman and Company: New York, NY, USA, 1995; p. 422.
7. Li, Y.; Mu, M.Y.; Yuan, L.; Zeng, B.M.; Lin, S.D. Challenges in the early diagnosis of patients with acute liver failure induced by amatoxin poisoning two case reports. *Medicine* **2018**, *97*, e11288. [CrossRef] [PubMed]
8. Costantino, D. Mushroom poisoning in Italy. *Int. J. Artif Organs* **1978**, *1*, 257–259. [CrossRef]

9. Diaz, J.H. Amatoxin-containing mushroom poisonings: Species, toxidromes, treatments, and outcomes. *Wilderness Environ. Med.* **2018**, *29*, 111–118. [CrossRef]
10. Koksal, O.; Armagan, E.; Inal, T.; Ayyildiz, T.; Oner, N. A 4 year retrospective analysis of our patients with mushroom poisoning. *Hong Kong J. Emerg. Med.* **2013**, *20*, 105–110. [CrossRef]
11. Ward, J.; Kapadia, K.; Brush, E.; Salhanick, S.D. Amatoxin poisoning: Case reports and review of current therapies. *J. Emerg. Med.* **2013**, *44*, 116–121. [CrossRef]
12. Jaeger, A.; Jehl, F.; Flesch, F.; Sauder, P.; Kopferschmitt, J. Kinetics of amatoxins in human poisoning—Therapeutic implications. *J. Toxicol. Clin. Toxicol.* **1993**, *31*, 63–80. [CrossRef]
13. Butera, R.; Locatelli, C.; Coccini, T.; Manzo, L. Diagnostic accuracy of urinary amanitin in suspected mushroom poisoning: A pilot study. *J. Toxicol. Clin. Toxicol.* **2004**, *42*, 901–912. [CrossRef] [PubMed]
14. Sun, J.; Niu, Y.M.; Zhang, Y.T.; Li, H.J.; Yin, Y.; Zhang, Y.Z.; Ma, P.B.; Zhou, J.; Huang, L.; Zhang, H.S.; et al. Toxicity and toxicokinetics of *Amanita exitialis* in beagle dogs. *Toxicon* **2018**, *143*, 59–67. [CrossRef] [PubMed]
15. Morel, S.; Fons, F.; Rapior, S.; Dubois, V.; Vitou, M.; Portet, K.; Dore, J.-C.; Poucheret, P. Decision-making for the detection of amatoxin poisoning: A comparative study of standard analytical methods. *Cryptogam. Mycol.* **2016**, *37*, 217–239. [CrossRef]
16. Broussard, C.N.; Aggarwal, A.; Lacey, S.R.; Post, A.B.; Gramlich, T.; Henderson, J.M.; Younossi, Z.M. Mushroom poisoning—From diarrhea to liver transplantation. *Am. J. Gastroenterol.* **2001**, *96*, 3195–3198. [PubMed]
17. Abbott, N.L.; Hill, K.L.; Garrett, A.; Carter, M.D.; Hamelin, E.I.; Johnson, R.C. Detection of alpha-, beta, and gamma-amanitin in urine by LC-MS/MS using $^{15}N_{10}$-alpha-amanitin as the internal standard. *Toxicon* **2018**, *152*, 71–77. [CrossRef]
18. Zhang, S.; Zhao, Y.F.; Li, H.J.; Zhou, S.; Chen, D.W.; Zhang, Y.Z.; Yao, Q.M.; Sun, C.Y. A simple and high-throughput analysis of amatoxins and phallotoxins in human plasma, serum and urine using UPLC-MS/MS combined with prime HLB elution platform. *Toxins* **2016**, *8*, 128. [CrossRef]
19. Leite, M.; Freitas, A.; Azul, A.M.; Barbosa, J.; Costa, S.; Ramos, F. Development, optimization and application of an analytical methodology by ultra performance liquid chromatography-tandem mass spectrometry for determination of amanitins in urine and liver samples. *Anal. Chim. Acta* **2013**, *799*, 77–87. [CrossRef]
20. Filigenzi, M.S.; Poppenga, R.H.; Tiwary, A.K.; Puschner, B. Determination of alpha-amanitin in serum and liver by multistage linear ion trap mass spectrometry. *J. Agric. Food Chem.* **2007**, *55*, 2784–2790. [CrossRef]
21. Maurer, H.H.; Schmitt, C.J.; Weber, A.A.; Kraemer, T. Validated electrospray liquid chromatographic-mass spectrometric assay for the determination of the mushroom toxins alpha- and beta-amanitin in urine after immunoaffinity extraction. *J. Chromatogr. B* **2000**, *748*, 125–135. [CrossRef]
22. Gicquel, T.; Lepage, S.; Fradin, M.; Tribut, O.; Duretz, B.; Morel, I. Amatoxins (alpha- and beta-amanitin) and phallotoxin (phalloidin) analyses in urines using high-resolution accurate mass LC-MS technology. *J. Anal. Toxicol.* **2014**, *38*, 335–340. [CrossRef]
23. Defendenti, C.; Bonacina, E.; Mauroni, M.; Gelosa, L. Validation of a high performance liquid chromatographic method for alpha amanitin determination in urine. *Forensic Sci. Int.* **1998**, *92*, 59–68. [CrossRef]
24. Abuknesha, R.A.; Maragkou, A. A highly sensitive and specific enzyme immunoassay for detection of beta-amanitin in biological fluids. *Anal. Bioanal. Chem* **2004**, *379*, 853–860. [CrossRef] [PubMed]
25. Bühlmann Laboratories, A.G. *Amanitin ELISA Product Insert*; Catalog# ek-am1; Bühlmann Laboratories AG: Schönenbuch, Switzerland, 2013.
26. Bever, C.S.; Adams, C.A.; Hnasko, R.M.; Cheng, L.W.; Stanker, L.H. Development and application of a lateral flow immunoassay for the rapid and sensitive detection of lethal amatoxins from mushrooms. *Toxins* **2020**, under review.
27. Enjalbert, F.; Gallion, C.; Jehl, F.; Monteil, H.; Faulstich, H. Amatoxins and phallotoxins in *Amanita* species-high-performance liquid-chromatographic determination. *Mycologia* **1993**, *85*, 579–584. [CrossRef]
28. Yilmaz, I.; Kaya, E.; Sinirlioglu, Z.A.; Bayram, R.; Surmen, M.G.; Colakoglu, S. Clinical importance of toxin concentration in *Amanita verna* mushroom. *Toxicon* **2014**, *87*, 68–75. [CrossRef] [PubMed]
29. Staack, R.F.; Maurer, H.H. New Bühlmann ELISA for determination of amanitins in urine—Are there false positive results due to interferences with urine matrix, drugs or their metabolites? *Toxichem. Krimtech.* **2001**, *68*, 68–71.
30. Cope, R.B. Mushroom poisoning in dogs. *Vet. Med.* **2007**, *102*, 95–100.

31. Puschner, B.; Wegenast, C. Mushroom poisoning cases in dogs and cats: Diagnosis and treatment of hepatotoxic, neurotoxic, gastroenterotoxic, nephrotoxic, and muscarinic mushrooms. *Vet. Clin. North Am. Small Anim. Pract.* **2018**, *48*, 1053–1067. [CrossRef]
32. Bever, C.S.; Barnych, B.; Hnasko, R.; Cheng, L.W.; Stanker, L.H. A new conjugation method used for the development of an immunoassay for the detection of amanitin, a deadly mushroom toxin. *Toxins* **2018**, *10*, 265. [CrossRef]
33. Bever, C.S.; Hnasko, R.M.; Cheng, L.W.; Stanker, L.H. A rapid extraction method combined with a monoclonal antibody-based immunoassay for the detection of amatoxins. *Toxins* **2019**, *11*, 724. [CrossRef]

toxins

MDPI

Article

Critical Comparison of Analytical Performances of Two Immunoassay Methods for Rapid Detection of Aflatoxin M_1 in Milk

Ivan Pecorelli [1], Natascia Guarducci [2], Cristoph von Holst [3], Rita Bibi [1], Michelangelo Pascale [2], Biancamaria Ciasca [2], Antonio F. Logrieco [2] and Veronica M. T. Lattanzio [2,*]

1 Istituto Zooprofilattico Sperimentale dell'Umbria e delle Marche "Togo Rosati", Via Salvemini 1, 06126 Perugia, Italy; i.pecorelli@izsum.it (I.P.); r.bibi@izsum.it (R.B.)
2 Institute of Sciences of Food Production, National Research Council of Italy, via Amendola 122/O, 70126 Bari, Italy; natascia.guarducci@gmail.com (N.G.); michelangelo.pascale@ispa.cnr.it (M.P.); biancamaria.ciasca@ispa.cnr.it (B.C.); antonio.logrieco@ispa.cnr.it (A.F.L.)
3 European Commission, Joint Research Centre (JRC), Retieseweg 111, 2440 Geel, Belgium; christoph.von-holst@ec.europa.eu
* Correspondence: veronica.lattanzio@ispa.cnr.it

Received: 26 March 2020; Accepted: 21 April 2020; Published: 22 April 2020

Abstract: Aflatoxin B_1 (AFB_1) is a secondary metabolite produced by some *Aspergillus spp.* fungi affecting many crops and feed materials. Aflatoxin M_1 (AFM_1), the 4-hydroxylated metabolite of AFB_1, is the main AFB_1-related compound present in milk, and it is categorized by the International Agency for Research on Cancer (IARC) as a "group 1 human carcinogen". The aim of this work was to evaluate and compare the analytical performances of two commercial immunoassays widely applied for the detection of AFM_1 in milk, namely strip test immunoassay and enzyme linked immunosorbent assay (ELISA). Assay validation included samples at AFM_1 levels of 25, 50, 75 ng/kg and blank samples (AFM_1 < 0.5 ng/kg). With respect to a screening target concentration (STC) of 50 ng/kg the two assays showed cut-off values of 37.7 ng/kg and 47.5 ng/kg for strip test and ELISA, respectively, a false suspect rate for blanks <0.1% (for both assays) and a false negative rate for samples containing AFM_1 at levels higher than STC, of 0.4% (for both assays). The intermediate precision (RSD_{ip}) was <32% for the strip test and <15% for the ELISA. Method verification through long-term intra-laboratory quality control (QC) measurements confirmed the results from the validation study. Furthermore, a satisfactory correlation of the results obtained with both immunoassays and the AOAC Official Method 2000.08 was obtained for the analysis of cow milk samples naturally contaminated with AFM_1 at levels within "not detected" (< 0.5 ng/kg) and 50 ng/kg. Finally, the extension of the scope of the strip test method to goat and sheep milk was evaluated by applying the experimental design foreseen in the EU regulation.

Keywords: Aflatoxin M_1; milk; strip test immunoassay; ELISA; method validation

Key Contribution: The study provides insights about the process of evaluating and comparing the performance profile of rapid methods currently applied for AFM_1 screening in milk. Addressing EU official guidelines, the fitness for purpose of strip test and ELISA based methods was evaluated, including verification of method performances through long-term quality control measurements and comparison with the AOAC reference method.

1. Introduction

Aflatoxins are mycotoxins found in four main chemical structures: aflatoxin B_1 (AFB_1), B_2 (AFB_2), G_1 (AFG_1) and G_2 (AFG_2); they can occur in a wide range of crops, including the major staple cereals (e.g., maize), edible nuts and legumes and their products. The main fungal producers of aflatoxins are *Aspergillus flavus* which produces mainly AFB_1, AFB_2 and *Aspergillus parasiticus*, which produces all four forms. Contamination can occur before or after harvest or both. In general, AFB_1 occurs at the highest levels compared to the others, and is the most toxic and a potent carcinogen [1,2]. AFB_1 is converted into its hydroxylated metabolite (AFM_1) by the liver enzymes of lactating animals [3]. This toxin, like the parent compound, has been categorized by the International Agency for Research on Cancer (IARC) as a group 1 toxin, a human carcinogen [2]. Due to their carcinogenity, the aflatoxins uptake through contaminated food consumption should be as low as possible, therefore the aflatoxin legislation is intended to implement the ALARA principle (As Low As Reasonably Achievable) and no threshold limit concerning the tolerable daily intake in humans has been established [4].

Evidences of aflatoxins carry over in milk, edible animal tissues and eggs have been reported, however, among foods of animal origin, milk represents the main source of human exposure to AFM_1, which is the only mycotoxin which has regulatory limits in milk [4–6]. There is evidence of AFM_1 occurrence in cow milk, but also in milk produced by other ruminants, such as buffalo, goat, sheep and camel [5]. The occurrence of AFM_1 has been reported in various locations worldwide. Overall, the incidence of AFM_1 in milk samples and milk products is relatively low in European countries, whereas data from Asian countries like China, Thailand and Taiwan show AFM_1 occurrence in up to 100% of samples [7]. AFM_1 is heat stable and processing or storage conditions are ineffective in reducing its concentration in milk and milk products [8–10]. Several factors may affect the AFM_1 contamination of milk, such as environmental conditions, different farming and feeding practices, as well as the quality and safety control systems put in place by food/feed business operators (FBO) [9,11]. The presence of AFM_1 in milk can be therefore considered as an indicator of maize chain vulnerability to fungal contamination [12].

Nowadays, there is an increasing concern for the impact of climate changes (temperature, humidity, rainfall and carbon dioxide production) on fungal behavior and consequently on aflatoxins production [11]. The application of predictive models has already given an indication of the potential increasing contamination by aflatoxins in Europe as consequence of climate changes [13]. Furthermore, a recent study using a full chain modeling approach to predict the impacts of climate change on AFB_1 production in maize and its consequences on AFM_1 contamination in dairy cow's milk, showed that, in the investigated scenario (i.e., Ukrainian maize), AFM_1 contamination in milk is expected to be comparable or to increase in future climate scenarios [14]. Therefore, according to EFSA definition, the presence of AFM_1 in milk may be considered as an "emerging risk", being a known risk for which an increasing and unpredictable pattern of exposure risk is foreseen [15].

Approximately 60 countries have already established regulatory limits for AFM_1 in milk and dairy products [16]. In the EU, the maximum permitted levels for AFM_1 have been set for consumable milk (50 ng/kg) [17]. In addition, an alert threshold level of 40 ng/kg calling for action is considered in some EU member states [18]. A maximum permitted level of 500 ng/kg of AFM_1 in milk has been established by the US-FDA (United States Food and Drug Administration) [19] and by the Codex Alimentarius [20]. This is also the harmonized MERCOSUR limit applied in Latin America [16,21] and in several Asian countries [16].

With the publication of the General Food Law (GFL) [22] the European Union has made a new legal framework laying down the principles, obligations and definitions that apply in the field of food safety. A general principle of the GFL is that FBOs have the primary responsibility for food and feed safety. To this purpose, FBOs must implement a food safety management system, based on the hazard analysis and critical control point (HACCP) principles. Regulatory limits therefore have a strong impact on contracts and procedural guidelines in the dairy industry and, as consequence, on the number of controls needed to verify milk compliance with maximum permitted levels, which may affect production costs.

A wide range of methods for the detection of AFM_1 in milk and dairy products is currently available, however, achieving key analytical performances, such as sensitivity, precision and reliability, suitable to enforce regulatory limits in the low ng/kg range, is still quite challenging [23]. Screening tests can play an important role within the safety monitoring, allowing rapid decision making and interventions, also affecting the final price of food products. Nowadays, screening tests based on immunochromatographic assays such as dipstick or lateral flow devices, and enzyme linked immunosorbent assays (ELISA) represent the most common formats in the market [24–26]. To support FBOs in selecting the most appropriate test in relation to the intended scope, internationally recognized guidelines for screening test performance verification have been made available for instance by the AOAC Research Institute (Performance Tested Methods^{SM}) and USDA-GIPSA (Performance Verified Rapid Test), whereas at European level, such guidelines are set in the Commission Decision 657/2002/EC [27] and in the Commission Regulation 2014/519/UE [28], which is specifically devoted to mycotoxin screening methods.

The aim of this work was to evaluate and compare the analytical performances of two commercial immunoassays (strip test immunoassay and ELISA) widely applied for the detection of AFM_1 in milk. For this purpose, the Commission Regulation 2014/519/UE [28] was taken into consideration as guidance document. Analytical performances, such as precision profile, cut-off value, false positive and false negative rates were evaluated for each assay by single laboratory validation, whereas a verification of the results from the validation study was performed based on long-term intra-laboratory quality control (QC) data. Correlation of the results obtained with the rapid immunoassays and the AOAC Official Method 2000.08 was evaluated for the analysis of naturally contaminated cow milk samples. Finally, the extension of the scope of the strip test method to goat and sheep milk was evaluated by applying the experimental design foreseen in the EU regulation [28].

2. Results and Discussion

The experimental design to evaluate analytical performances of strip test immunoassay and ELISA comprised the following steps, which were carried out in parallel for the two assays: i) single laboratory validation study to evaluate precision, cut-off value, false suspect and false negative rates (milk samples fortified by AFM_1 were used at this stage); ii) verification of cut-off and precision values by long-term intra-laboratory QC study (a QC cow milk sample spiked at 50 ng/kg was used at this stage); iii) evaluation of results correlation between rapid immunoassays and AOAC Official Method 2000.08 (a set of naturally contaminated cow milk samples was used for this purpose). Data obtained for each step are described and discussed in the following.

2.1. Validation Results

Validation experiments were performed according to the experimental design described in Section 4.6. The screening target concentration (STC) value was 50 ng/kg. Other tested mass fractions values were: blank ($AFM_1 \leq 0.5$ ng/kg), 25 ng/kg (50% of the STC), 75 ng/kg (150% STC). The same sample set was analyzed by the ELISA and the strip test. Results obtained from the 24 measurements performed for each validation level were taken as basis for the calculation of validation parameters precision, cut-off value, false positive and false negative rate. The overall results of the statistical assessment are shown in Table 1.

Table 1. Analytical performances of the strip test immunoassay and enzyme linked immunosorbent assay (ELISA), as resulted by validation experiments.

Test Sample		Strip Test	ELISA
Blank ($AFM_1 \leq 0.5$ ng/kg)	Mean response (ng/kg)	4.2	8.2
	RSD_r (%)[1]	93	16
	RSD_{ip} (%)[2]	140	33
	False suspect rate (%)	<0.1	<0.1
50% STC (25 ng/kg)	Mean response (ng/kg)	23.2	30.7
	RSD_r (%)	26	5
	RSD_{ip} (%)	32	5
	False suspect rate (%)	3	<0.1
STC (50 ng/kg)	Mean response (ng/kg)	53.1	57.9
	RSD_r (%)	12	5
	RSD_{ip} (%)	17	10
150% STC (75 ng/kg)	Mean response (ng/kg)	85.6	105
	RSD_r (%)	16	15
	RSD_{ip} (%)	19	15
	False negative rate (%)	0.4	0.1
Cut-off value (ng/kg)		37.7	47.5

[1] RSD_r relative standard deviation of the repeatability; [2] RSD_{ip} relative standard deviation for intermediate precision.

First, precision data were calculated for all tested concentrations. Specifically, RSD_{ip} (intermediate precision) values of 32% (strip test) and 5% (ELISA) were obtained for samples contaminated at 25 ng/kg, values of 17% (strip test) and 10% (ELISA) at 50 ng/kg, 19% (strip test) and 15% (ELISA) at 75 ng/kg. Repeatability values (RSD_r) were lower than 26% in all cases. Comparable values were obtained for the two tests at STC and above STC, whereas at 50% STC (25 ng/kg) lower intermediate precision values where obtained for ELISA. This could be partially explained by the fact that ELISA was working at a level five times higher than its limit of detection (LOD, 5 ng/kg, see Section 4.4), whereas the strip test was working at its LOD (25 ng/kg see Section 4.3).

With respect to the blank samples, a high relative standard deviation of the strip test response was observed. This could be mainly explained by the fact that for the strip test assay analytical signal values below a certain fixed limit, which is set by the manufacturer, are reported as "zero concentration", whether that is true or not. This led to a high number of "zero concentration" values in blank samples generating a high standard deviation. However, in the following, it will be shown that, notwithstanding this high value, an acceptable low rate of false suspect results for the blank samples was obtained anyway, due to the good separation of test responses for blank and contaminated samples.

Overall, the obtained precision values indicated an acceptable robustness of the two test methods, also taking into consideration the very low target levels of AFM_1 considered for validation.

Once intermediate precision data were available, it was possible to calculate the cut-off values. According to European legislation [28], this value is defined as the response (AFM_1 mass fraction) obtained with the screening method, "above which the sample is classified as suspect", with a false negative rate of 5%. The calculated cut-off values were 37.7 ng/kg for the strip test and 47.5 ng/kg for the ELISA test, respectively. In both cases, the assay sensitivity was considered satisfactory for assessing milk contamination at levels encompassing the EU maximum limits.

Based on the cut-off values, the rate of false suspect results was estimated for samples containing AFM_1 below the STC. Specifically, for samples contaminated at 50% STC (25 ng/kg) the false suspect rate was 3% for the strip test and < 0.1% for the ELISA test. Finally, the false negative rate for samples contaminated at levels above the STC (75 ng/kg in the present case) resulted to be 0.4% for strip test and 0.1% for ELISA. In both cases, the acceptability criterion of maximum 5% false negative rate was met.

The overall results indicated satisfactory kits reliability in discriminating samples contaminated at different AFM_1 levels set in a very narrow working range (from ≤ 0.5 to 75 ng/kg), encompassing EU regulatory limits.

In addition, the method fitness for purpose of evaluating milk contamination at the alert threshold of 40 ng/kg was evaluated by analyzing 20 contaminated samples from two different farms. The obtained average responses were 40.0 ng/kg for the strip test and 40.2 for the ELISA, with relative standard deviation (RSD_{ip}) of 9.8% and 5.8%, respectively. The resulting cut-off levels were 33.2 ng/kg and 36.1 ng/kg. No false suspect samples resulted for blanks with respect to these cut-off values. These data demonstrated the fitness for purpose of the two tested kits in evaluating compliance of milk samples with respect to the alert threshold of 40 ng/kg.

2.2. *Verification of Method Performances through Quality Control Data*

Verification of method performances was carried out through long-term intra-laboratory QC measurements over a period of 12 months (see Section 4.6). The results of the validation and the verification study were compared in terms of precision, recovery rates and cut-off values, as shown in Table 2. The cut-off values calculated by QC data matched very well with those obtained by validation data. Moreover, the data from the validation study as well as from the QC exercise revealed comparable values for the precision and the recovery rate, thus demonstrating sufficient ruggedness of both methods over the time and different production lots.

Table 2. Verification of strip test and ELISA method performances though quality control (QC) data.

Method Performances		Strip Test		ELISA	
		Single Lab validation	QC	Single Lab validation	QC
STC (50 ng/kg)	Mean response (ng/kg)	53.1	53.5	57.9	52.8
	Relative recovery rate (%)	106	107	116	106
	RSD_{ip} (%) [1]	17	12	10	9.2
	Cut-off value (ng/kg)	37.7	42.7	47.5	44.7

[1] RSD_{ip} relative standard deviation for intermediate precision.

2.3. *Analysis of Naturally Contaminated Samples*

The trueness of data generated by the two screening methods was evaluated by comparing them with results obtained by the reference AOAC Official Method 2000.08 on a set of raw cow milk samples naturally contaminated in the range n.d. (≤0.5 ng/kg) – 50 ng/kg AFM_1.

Results are depicted in Figure 1. The two test kits performed in a similar way, and in both cases a satisfactory correlation was observed, with results provided by the reference method (r = 0.923 and slope = 0.84 for strip test vs HPLC and r = 0.924 and slope = 1.05 for ELISA vs HPLC). Irrespective of a slight overestimation of the AFM_1 content in some of the blank samples (HPLC result ≤ 0.5 ng/kg), both immunoassays returned values lower than 14 ng/kg, confirming the absence of false suspect results.

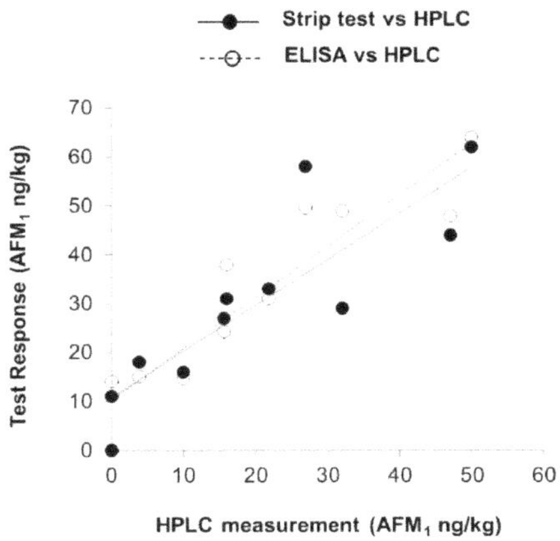

Figure 1. Correlation between results (AFM_1 mass fraction, ng/kg) obtained by strip test or ELISA and the HPLC analysis performed according to the AOAC Official Method 2000.08.

2.4. Extension of Scope of the Method to Other Commodities

Finally, the extension of the scope of the strip test method to goat and sheep milk was evaluated by applying the experimental design foreseen in the EU regulation. The regulation foresees that "as long as the new commodity belongs to a commodity group ("milk" in the present case) for which an initial validation has already been performed, a minimum of 10 homogeneous negative control and 10 homogeneous positive control (at STC) samples shall be analyzed under intermediate precision conditions. The positive control samples shall all be above the cut-off value as calculated in validation experiments.

For these purposes, first specific calibration curves (bar codes) were generated for strip test analysis of raw goat and raw sheep milk. Then 10 blank (negative) samples and 10 samples contaminated by AFM_1 at 50 ng/kg were analyzed for each milk type. An additional sample set containing AFM_1 at 25 ng/kg was also included. Results are reported in Figure 2. In both cases negative samples were correctly classified as below the cut-off. No false suspect was reported. In addition, samples contaminated at 50% STC (25 ng/kg) were all correctly classified as below the cut-off (Table 1) and no false suspect was reported. All samples contaminated at 50 ng/kg (STC) were correctly classified above the cut-off.

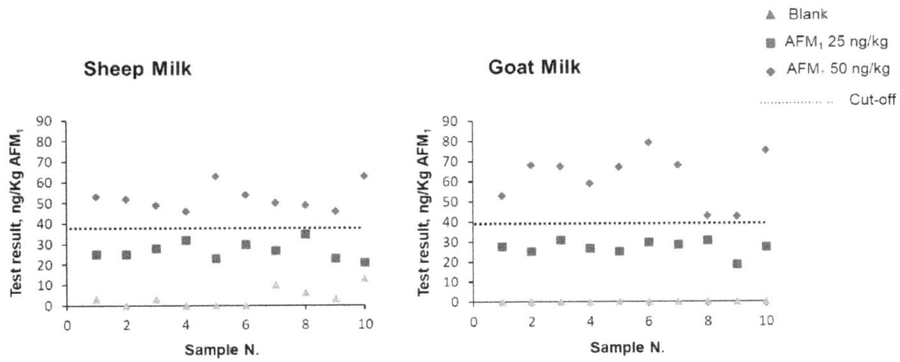

Figure 2. Results of strip test analysis of blank sheep and goat milk samples and samples contaminated with 25 and 50 ng/kg AFM_1.

The obtained data showed the applicability of the strip test immunoassay to goat and sheep milk provided that a specific calibration curve was used.

2.5. Fitness for Purpose of the Validated Immunoassays

Validation experiments returned, for both immunoassays, fit for purpose analytical performances such as cut-off values (37.7 ng/kg and 47.5 ng/kg for strip test and ELISA respectively), false suspect rate for blanks (<0.1% for both assays) and false negative rate (<0.4% (for both assays). Both assays showed an intermediate precision at STC (50 ng/kg) <17% either in validation and QC measurements. However, besides analytical performances, when choosing a method for rapid mycotoxin screening the concept of fitness for purpose also includes some practical parameters. Factors such as the time needed for analysis, the skills or level of education of the user of the method and the place where the analysis needs to be carried out are generally taken into consideration by the end users. A more comprehensive comparison of performances of mycotoxin screening tests can be found in Lattanzio et al. [29]. In the present case, the total analytical time for strip test assay was about 10 min and the use of the incubator, as well as the portable reader, made it suitable for on farm use. The ELISA involved more steps, a basic laboratory equipment and more time (approx 80 min). On the other hand, ELISA tests allow to handle up to 48 samples simultaneously (including calibrants and QC samples), while the strip test foresees only one sample per analysis/strip. ELISA can be therefore more efficient when a large number of (sub)samples need to be analyzed in a short period of time. On the other hand, when applied in routine by experienced technicians, strip testing can be stacked to process multiple samples in a relatively short period of time, by processing 10 to 15 samples 1 min apart. Finally, concerning method transferability to unskilled personnel, the strip test appears easier to be applied by low experienced technicians, not only because the analytical protocol is less laborious, but also because the automatic calibration via QR code uploading. In principle both platforms are potentially suitable for multiplexing [30–32].

3. Conclusions

Analytical performances and fitness for purpose of two commercial immunoassays widely applied for the detection of AFM_1 in milk (strip test and ELISA) were evaluated, according to guidelines set in Regulation 519/2014/EU. Both assays showed satisfactory performances in terms of precision, recovery rates, false positive and false negative rates. In addition, the method performance profiles of the two methods obtained in the validation study could be verified by long-term intra-laboratory QC data. A good correlation between the results provided by the validated assays and the AOAC reference method was observed when analyzing naturally contaminated samples. The extension of the scope of the strip test method to goat and sheep milk was successfully evaluated by applying the experimental design foreseen in the EU regulation.

4. Materials and Methods

4.1. Materials

Methanol and acetonitrile (HPLC grade) were purchased from Mallinckrodt Baker (Milan, Italy). Ultrapure water was produced by using a Millipore Milli-Q system (Millipore, Bedford, MA, USA). AflaM1 Test immunoaffinity columns were from VICAM (A Waters Business, Milford, MA, USA). Paper Filters (Whatman 4) were obtained from Whatman International Ltd. (Maidstone, UK). Standard aflatoxin M_1 (acetonitrile solution 10 μg/mL) was purchased from Supelco (Bellefonte, PA, USA).

4.2. Milk Samples

Milk samples (cow, sheep, goat) were collected from farm bulk tanks from Italian farmers in the time span 2018–2019. After collection samples were stored at 4 °C and analyzed within 48 h.

Fortified milk samples to be used for validation (Section 4.6) were prepared as follows. A spiking solution containing AFM_1 at 1 µg/mL was prepared by diluting 10 times the standard AFM_1 solution. Then 25 µL of spiking solution were added to 50 mL of milk to prepare a mother solution at 500 ng/kg AFM_1. The mother solution was diluted by appropriate volumes of milk to obtain contaminated samples at 75, 50, 40 and 25 ng/kg.

4.3. Strip Test Immunoassay

The strip test (AFLAM1-V™), incubator, and a photometric reader (Vertu Reader) were from VICAM (A Waters Business, Milford, MA, USA). The strip test format is based on an indirect competitive immunoassay. Line intensities developed on the strip membrane (test line and control line) are measured using the photometric reader. The test response is the ratio between the signal intensity of the test line and that of the control line and it is converted into toxin concentration through a lot specific calibration curve.

Strip test analyses of milk samples were performed as follows. Two hundred microliters of cold milk were pipetted into the reagent vial. After vortex mixing (3 times x 5 sec) the vial was placed into the incubator set at 40 °C and the strip was inserted into it. The sample was allowed to migrate onto the strip for to 10 min, then the strip was placed into the reader holder for result reading. The lot specific calibration curve was uploaded onto the reader system by using the corresponding barcode provided by the supplier. The calibration curve was generated by spiking uncontaminated milk (cow, sheep or goat milk) at seven AFM_1 levels over the range 0–800 ng/kg, performing triplicate measurements for each calibration level. The limit of detection declared by the supplier was 25 ng/kg.

4.4. Enzyme Linked Immunosorbent Assay

The ELISA kit (I'Screen AflaM1) was from Eurofins Technologies (Budapest, Hungary). Calibration standards, conjugate, antibody and substrate/chromogen solutions were provided in the kit. The plate reader was Multiskan MS Plus MK II ELISA reader from Labsystems (Helsinki, Finland).

Samples were analyzed as follows. One hundred microliters of milk (or calibrant solution) were transferred into the well and incubated for 45 min at room temperature. After discarding the liquid by turning the plate upside down, the wells were filled completely with the working wash solution. Then the liquid was poured out from wells and the remaining drops were removed by tapping the microplate upside down against adsorbent paper. This washing sequence was repeated four times. Then 100 µL of AFM_1-enzyme conjugate solution were added and incubated for 15 min. After discarding the liquid, the wells were washed four times, according to the above described procedure. Then, 100 µL of substrate solution were added and incubated for 15 min for color development. Finally, 50 µL of stop solution were added. Result were read measuring the absorbance at 450 nm. The limit of detection declared by the supplier was 5 ng/kg.

4.5. Reference Method (AOAC Official Method 2000.08)

Screening for blank samples to be used for validation experiments and strip test calibration curve generation, and analysis of naturally contaminated samples for method comparison purposes were performed, according to the AOAC Official Method 2000.08, with minor modifications. Briefly, milk samples (50 mL) were centrifuged at 2000× *g* to separate the fat. After discarding the upper thin fat layer, the sample was filtered through paper filter. The filtered sample (25 mL) was passed through the immunoaffinity column. The eluate was discarded and the column was washed twice with 10 mL distilled water. The toxin was eluted by 2 × 1 mL methanol. The eluate was collected and dried down under nitrogen stream. The residue was re-dissolved with 250 µL of a mixture water:acetonitrile (75:25 by vol).

HPLC-FD analyses were performed on an Agilent 1100 Series chromatographic system (Agilent Technologies, Palo Alto, CA, USA) equipped with a fluorometric detector (model 363), and the ChemStation data software (Agilent Technologies). The analytical column was a Zorbax SB-C18 Rapid

Resolution HT (4.6 mm × 50 mm, 1.8 µm) with corresponding in-line filter. The chromatographic separation was performed by a gradient elution using water (solvent A) and acetonitrile (solvent B). The flow rate of the mobile phase was 1 mL/min. The injection volume was 50 µL. The column was kept at a temperature of 40 °C; the excitation wavelength was set at 365 nm and the emission wavelength was set at 435 nm. The detection limit was 0.5 ng/kg.

4.6. Validation Design and Verification Study via Qualitity Control (QC) Measurements

The single laboratory validation study was designed to fulfil the specifications established in Commision Regulation (EU) 519/2014, in terms of minimum sample set and minimum number of validation levels.

Measurements were distributed in two different days (instead of 5 as suggested in the regulation, due to the limited stability of milk samples). Milk samples were from three different farms. In addition, each sample was analyzed in quadruplicate each day under repeatability conditions. The design resulted in 12 independent analysis per day and 24 measurements in total per each validation level.

The screening target concentration (STC) was set at the EU maximum permitted level of 50 ng/kg. The selected validation levels were four: blank (i.e., <0.5 ng/kg), and samples spiked by 25 ng/kg (50% STC), 50 ng/kg (STC), 75 ng/kg (150% STC) of AFM_1. An additional sample set containing AFM_1 at 40 ng/kg was included to evaluate the fitness for purpose of the two tested methods in high risk periods when it is recommended to set the alert threshold (STC) at 40 ng/kg [18].

The results of analysis were then subjected to statistical assessment to calculate validation parameters as described in the following.

Verification of method performances was carried out through long-term intra-laboratory QC measurements. For both assays, 50 measurements of the QC material, i.e., raw cow milk spiked at STC (50 ng/kg), were spread over a period of 12 months. Moreover, 4 different kit lots were used for the strip test and 6 different lots for the ELISA test, thus including additional factors in the verification study, which may have an impact on the result of analysis. Finally, the results of these analysis were taken as a basis for the calculation of the cut-off values, precision and recovery rates.

4.6.1. Precision

To evaluate the precision profile of the method, data generated each validation level were subjected to analysis of variance (ANOVA). When applying this model, as defined by ISO 5725 [33], the measured test response Y_{ijk} (the AFM_1 mass fraction in the present case) is defined as the true value (TV) plus the contribution of 3 components:

$$Y_{ijk} = TV + D_i + M_{ij} + R_{ijk} \tag{1}$$

where D_i is the between-day variability, M_{ij} is the between-matrix (milk batches from different farms) variability, and R_{ijk} is the within-day variability. The within-day variability gives the precision under repeatability conditions, whereas the sum of all components gives the intermediate precision. The statistical assessment was done with the software package MINITAB™ Statistical Software for Windows (Version 15).

4.6.2. Cut-Off Value

The measured levels (ng/kg) of samples containing AFM_1 at STC were taken as basis for the calculation of the cut-off value. According to Regulation 519/2014/EU the following equation was used:

$$Cut\ off = R_{STC} - tvalue_{(0.05)} \times SD_{STC} \tag{2}$$

where the R_{STC} is the mean level of AFM_1 (ng/kg) calculated from all 24 experiments performed on samples containing AFM_1 at STC, SD_{STC} is the corresponding standard deviation of intermediate precision as defined in the previous paragraph, and $tvalue_{(0.05)}$ is the one tailed t value for a rate of false negative results of 5%.

4.6.3. False Suspect and False Negative Rate

Using the cut-off value and the results from the analysis of negative samples the rate of false suspect results was estimated by first calculating the t-value as follows:

$$tvalue = \frac{\left(cut\,off - mean_{neg}\right)}{SD_{neg}} \tag{3}$$

where $mean_{neg}$ is the mean value of the results obtained from the 24 experiments on the negative samples and SD_{neg} is the corresponding standard deviation of intermediate precision.

From the obtained t-value, based on the degrees of freedom calculated from the number of experiments (23 in the present case), the false suspect rate results (probability) for a one tailed distribution was calculated using the spread sheet function "TDIST" from Microsoft Excel.

The false suspect rate for samples containing AFM_1 at 50% STC was calculated by applying the same procedure using the mean value of the results obtained from the 25 experiments on samples containing AFM_1 at 50% STC and the relevant standard deviation of intermediate precision.

Finally, the false negative rate for samples containing AFM_1 at levels above the STC was estimated by calculating the t value as specified here:

$$tvalue = \frac{(mean_{>STC} - cutoff)}{SD_{>STC}} \tag{4}$$

where $mean_{>STC}$ is the mean value of the results obtained from the experiments on the samples containing the analyte above the STC, cut-off is the value established as above, and $SD_{>STC}$ is the corresponding standard deviation of the intermediate precision. The probability corresponding to the calculated t value with a one-tailed distribution gives the rate of false negative results for the samples containing the analyte at levels higher than STC.

Author Contributions: Conceptualization, V.M.T.L. and C.v.H.; methodology, V.M.T.L. and I.P.; formal analysis and validation, N.G., R.B. and B.C.; statistics, C.v.H.; writing—original draft preparation, V.M.T.L., I.P.; writing—review and editing, supervision, V.M.T.L., M.P. and A.F.L. All authors have read and agreed to the published version of the manuscript.

Funding: This research received no external funding.

Conflicts of Interest: The authors declare no conflict of interest.

References

1. WHO-FAO. Evaluation of Certain Food Additives and Contaminants, Sixty-Eighth Report of the Joint FAO/WHO Expert Committee on Food Additives. WHO Technical Report Series 947. Available online: http://apps.who.int/iris/bitstream/10665/43870/1/9789241209472_eng.pdf (accessed on 20 April 2020).
2. International Agency for Research on Cancer. Monographs on the Evaluation of Carcinogenic Risks to Humans: Chemical Agents and Related Occupations. A Review of Human Carcinogens. Volume 100F. Available online: http://publications.iarc.fr/Book-And-Report-Series/Iarc-Monographs-On-The-Identification-Of-Carcinogenic-Hazards-To-Humans/Chemical-Agents-And-Related-Occupations-2012 (accessed on 20 April 2020).
3. Marchese, S.; Polo, A.; Velotto, S.; Costantini, S.; Severino, L. Aflatoxin B1 and M1: Biological properties and their involvement in cancer development. *Toxins* **2018**, *10*, 214. [CrossRef]
4. WHO-FAO. Safety Evaluation of Certain Contamimants in Food. Eighty-Third Meeting of the Joint FAO/WHO Expert Committee on Food Additives. WHO Technical Report Series 74. Available online: https://apps.who.int/iris/handle/10665/276868 (accessed on 17 April 2020).
5. European Food Safety Authority (EFSA). Opinion of the Scientific Panel on Contaminants in the Food Chain on a request from the Commission related to Aflatoxin B1 as undesirable substance in animal feed. *EFSA J.* **2004**, *39*, 1–27. [CrossRef]

6. Völkel, I.; Schröer-Merker, E.; Czerny, C.P. The carry-over of mycotoxins in products of animal origin with special regards to its implications for the European food safety legislation. *Food Nutr. Sci.* **2011**, *2*, 852–867 [CrossRef]
7. Becker-Algeri, T.A.; Castagnaro, D.; De Bortoli, K.; De Souza, C.; Drunkler, A.; Badiale-Furlong, E. Mycotoxins in bovine milk and dairy products: A review. *J. Food Sci.* **2016**, *81*, R544–R552. [CrossRef] [PubMed]
8. WHO-FAO. Safety Evaluation of Certain Mycotoxins in Food—Prepared by the 56th meeting of the JECFA (Joint FAO/WHO Expert Committee on Food Additives)—FAO Food and Nutrition Paper 74/WHO Foods Additives Series 47. Available online: http://www.inchem.org/documents/jecfa/jecmono/v47je01.htm (accessed on 20 April 2020).
9. Prandini, A.; Tansini, G.; Sigolo, S.; Filippi, L.; Laporta, M.; Piva, G. On the occurrence of aflatoxin M1 in milk and dairy products. *Food Chem. Toxicol.* **2009**, *47*, 984–991. [CrossRef]
10. Campagnollo, F.B.; Ganev, K.C.; Khaneghah, A.M.; Portela, J.B.; Cruz, A.G.; Granato, D.; Corassin, C.H. Oliveira, C.A.F.; Sant'Ana, A.S. The occurrence and effect of unit operations for dairy products processing on the fate of aflatoxin M1: A review. *Food Control* **2016**, *68*, 310–329. [CrossRef]
11. Miliĉeviĉ, D.; Petronijeviĉ, R.; Petroviĉ, Z.; Đjinoviĉ-Stojanoviĉ, J.; Jovanoviĉ, J.; Baltiĉ, T.; Janković, S. Impact of climate change on aflatoxin M1 contamination of raw milk with special focus on climate conditions in Serbia. *J. Sci. Food Agric.* **2019**, *99*, 5202–5210. [CrossRef]
12. Frazzoli, C.; Gherardi, P.; Saxena, N.; Belluzzi, G.; Mantovani, A. The hotspot for (global) one health in primary food production: Aflatoxin M1 in dairy products. *Front. Public Health* **2017**, *4*. [CrossRef]
13. Battilani, P.; Rossi, P.; Giorni, A.; Pietri, A.; Gualla, H.J.; van der Fels-Klerx, C.J.H.; Booij, A.; Moretti, A.; Logrieco, A.; Miglietta, F.; et al. Modelling, predicting and mapping the emergence of aflatoxins in cereals in the EU due to climate change. *EFSA Supporting Publ.* **2012**, *9*. [CrossRef]
14. Van der Fels-Klerx, H.J.; Vermeulen, L.C.; Gavai, A.K.; Liu, C. Climate change impacts on aflatoxin B1 in maize and aflatoxin M1 in milk: A case study of maize grown in Eastern Europe and imported to the Netherlands. *PLoS ONE* **2019**, *14*, e0218956. [CrossRef]
15. EFSA Definition and Description of "Emerging Risks". Available online: https://www.efsa.europa.eu/en/topics/topic/emerging-risks (accessed on 20 April 2020).
16. Food and Agriculture Organization of the United Nations. Worldwide Regulations for Mycotoxins in Food and Feed in 2003. Food and Nutrition Paper No. 81. Available online: http://www.fao.org/3/y5499e/y5499e00.htm (accessed on 20 April 2020).
17. European Commission. Commission Regulation (EC) No 401/2006 of 23 February 2006 laying down the methods of sampling and analysis for the official control of the levels of mycotoxins in foodstuffs. *Off. J. Eur. Commun.* **2006**, *L364*, 5–24.
18. Italian Ministry of Health. Nota del Ministero della Salute prot. n. 855 del 16/1/2013—Aflatossine. Extraordinary Operating Procedures for the Prevention and Management of the Risk of Aflatoxin Contamination in the Dairy Supply Chain and in the Production of Maize Destined for Human and Animal Nutrition, Following Extreme Weather Conditions. Available online: http://www.unionalimentari.com/website/law.aspx?id=1287&key=&fl= (accessed on 20 April 2020).
19. Food and Drug Administration. CPG Sec 527.400 Whole Milk, Lowfat Milk, Skim Milk—Aflatoxin M1. Available online: https://www.fda.gov/regulatory-information/search-fda-guidance-documents/cpg-sec-527400-whole-milk-lowfat-milk-skim-milk-aflatoxin-m1 (accessed on 20 April 2020).
20. Codex Alimentarius. General Standard for Contaminants and Toxins in Food and Feed (CODEX STAN 193-1995) Adopted in 1995 Revised in 1997, 2006, 2008, 2009 Amended in 2010, 2012, 2013, 2014, 2015. Available online: http://www.fao.org/fao-who-codexalimentarius/codex-texts/all-standards/en/ (accessed on 20 April 2020).
21. Agencia Nacional de Vigilancia Sanitaria (ANVISA), Brazil. Resoluçao RDC n_7, de 18 de Fevereirode 2011. Regulamento Tecnico Sobre Limites Maximos Tolerados (LMT) para Micotoxinas em Alimentos. Available online: http://bvsms.saude.gov.br/bvs/saudelegis/anvisa/2011/res0007_18_02_2011_rep.html (accessed on 20 April 2020).
22. European Commission. Commission Regulation (EC) No 178/2002 of the European Parliament and of the Council of 28 January 2002 laying down the general principles and requirements of food law, establishing the European Food Safety Authority and laying down procedures in matters of food safety. *Off. J. Eur. Commun.* **2002**, *L31*, 1–24.

23. Vaz, A.; Cabral Silva, A.C.; Rodrigues, P.; Venâncio, A. Detection methods for aflatoxin M1 in dairy products. *Microorganisms* **2020**, *8*, 246. [CrossRef] [PubMed]
24. Li, W.; Powers, S.; Dai, S.Y. Using commercial immunoassay kits for mycotoxins: 'joys and sorrows'? *World Mycotoxin J.* **2014**, *7*, 417–430. [CrossRef]
25. Matabaro, E.; Ishimwe, I.; Uwimbabazi, E.; Lee, B.H. Current immunoassay methods for the rapid detection of aflatoxin in milk and dairy products. *Compr. Rev. Food Sci. Food Saf.* **2019**, *16*, 808–820. [CrossRef]
26. Nolan, P.; Auer, S.; Spehar, A.; Elliott, C.T.; Campbell, K. Current trends in rapid tests for mycotoxins. *Food Addit. Contam. Part A* **2019**, *36*, 800–814. [CrossRef]
27. European Commission. Commission Decision (EC) No 657/2002 of 12 August 2002 implementing Council Directive 96/23/EC concerning the performance of analytical methods and the interpretation of results. *Off. J. Eur. Commun.* **2002**, *L221*, 8–36.
28. European Commission. Commission Regulation (EU) No 519/2014 of 16 May 2014 amending Regulation (EC) No 401/2006 as regards methods of sampling of large lots, spices and food supplements, performance criteria for T-2, HT-2 toxin and citrinin and screening methods of analysis. *Off. J. Eur. Commun.* **2014**, *L147*, 29–43.
29. Lattanzio, V.M.T.; Von Holst, C.; Lippolis, V.; De Girolamo, A.; Logrieco, A.F.; Mol, H.G.J.; Pascale, M. Evaluation of mycotoxin screening tests in a verification study involving first time users. *Toxins* **2019**, *11*, 129. [CrossRef]
30. Lattanzio, V.M.T.; Nivarlet, N.; Lippolis, V.; Della Gatta, S.; Huet, A.C.; Delahaut, P.; Granier, B.; Visconti, A. Multiplex dipstick immunoassay for semi-quantitative determination of Fusarium mycotoxins in cereals. *Anal. Chim. Acta* **2012**, *718*, 99–108. [CrossRef] [PubMed]
31. Maragos, C.M. Multiplexed biosensors for mycotoxins. *J. AOAC Int.* **2016**, *99*, 849–860. [CrossRef] [PubMed]
32. Plotan, M.; Devlin, R.; Porter, J.; Benchikh, M.; Rodríguez, M.L.; McConnell, R.I.; Fitzgerald, S.P. The use of biochip array technology for rapid multimycotoxin screening. *J. AOAC Int.* **2016**, *99*, 878–888. [CrossRef] [PubMed]
33. ISO. *Accuracy (Trueness and Precision) of Measurement Methods and Results—Part 3: Intermediate measures of the Precision of a Standard Measurement Method*; ISO 5725-3; ICS 03.120.30; ISO Standards: Geneva, Switzerland, 1994.

toxins

MDPI

Review

Two-Dimensional Layered Nanomaterial-Based Electrochemical Biosensors for Detecting Microbial Toxins

Zhuheng Li [1,2,†], Xiaotong Li [2,†], Minghong Jian [2], Girma Selale Geleta [2,3,*] and Zhenxin Wang [2,*]

1 Jilin Provincial Institute of Education, Changchun 130022, China; lizhuheng@126.com
2 State Key Laboratory of Electroanalytical Chemistry, Changchun Institute of Applied Chemistry, Chinese Academy of Science, Changchun 130022, China; lixiaotong@ciac.ac.cn (X.L.); mhjian@ciac.ac.cn (M.J.)
3 Department of Chemistry, College of Natural Sciences, Jimma University, Jimma 378, Ethiopia
* Correspondence: ggirma245@gmail.com (G.S.G.); wangzx@ciac.ac.cn (Z.W.); Tel.: +251-(0471)-111252 (G.S.G.); +86-(0431)-85262243 (Z.W.)
† These authors contributed equally.

Received: 4 November 2019; Accepted: 27 December 2019; Published: 31 December 2019

Abstract: Toxin detection is an important issue in numerous fields, such as agriculture/food safety, environmental monitoring, and homeland security. During the past two decades, nanotechnology has been extensively used to develop various biosensors for achieving fast, sensitive, selective and on-site analysis of toxins. In particular, the two dimensional layered (2D) nanomaterials (such as graphene and transition metal dichalcogenides (TMDs)) and their nanocomposites have been employed as labels and/or biosensing transducers to construct electrochemical biosensors for cost-effective detection of toxins with high sensitivity and specificity. This is because the 2D nanomaterials have good electrical conductivity and a large surface area with plenty of active groups for conjugating 2D nanomaterials with the antibodies and/or aptamers of the targeted toxins. Herein, we summarize recent developments in the application of 2D nanomaterial-based electrochemical biosensors for detecting toxins with a particular focus on microbial toxins including bacterial toxins, fungal toxins and algal toxins. The integration of 2D nanomaterials with some existing antibody/aptamer technologies into electrochemical biosensors has led to an unprecedented impact on improving the assaying performance of microbial toxins, and has shown great promise in public health and environmental protection.

Keywords: two dimensional layered nanomaterials; electrochemical biosensors; microbial toxin detection; antibodies; aptamers

Key Contribution: This review updates the construction strategies of electrochemical biosensors such as immunosensors and aptasensors for cost-effective determination of microbial toxins with high sensitivity given by rapidly developing two dimensional layered (2D) nanomaterial-based labels and substrates. We summarize the roles of 2D nanomaterials and their nanocomposites in the configuration of electrochemical biosensors, as well as the advantages they provide to the analyses, and address the major challenges and perspectives of these electrochemical biosensors for future commercialization.

1. Introduction

Two-dimensional layered (2D) nanomaterial (e.g., graphene and its derivatives, transition metal dichalcogenides (TMDs) and other layered nanosheets)-based electrochemical signal amplifications have great potential for improving both the sensitivity and selectivity of electrochemical biosensors because of their unique physical, chemical, and electrical properties [1–21]. Graphene is a single layer of densely packed carbon atoms with a benzene-ring structure, and is the first known 2D layered nanomaterial [22]. The unique properties of graphene, including its exceptional mechanical strength [23], extremely large surface area (2630 m^2/g) [24], very high thermal conductivity in the range of ~3080–5150 W mK^{-1} [25], high conductivity [26], good charge carrier mobility [27], and wide potential window [28], endow it with great applicability in the development of biosensors, and, in particular, electrochemical biosensors [1–5,12,22–24,29]. In addition, based on the molar ratio of carbon to oxygen (C/O), graphene can be roughly divided into two categories, graphene oxide (GO) or reduced graphene oxide (rGO). It is demonstrated that rGO has better electrical conductivity than GO. Because pure graphene lacks an intrinsic band gap and is limited by chemical modification, there is an increasing interest in synthesizing graphene derivatives/nanocomposites and graphene-like 2D nanomaterials. Among the graphene-like 2D nanomaterials, TMDs (e.g., molybdenum disulfide (MoS_2) and molybdenum selenide ($MoSe_2$)) show excellent physicochemical properties and remarkable biocompatibility, and also have significant attraction for the fabrication of electrochemical (bio)sensors [6,7,9,14,21,29]. Driven by their unprecedented properties, massive synthetic methods/protocols have been developed for preparing 2D nanomaterials and 2D nanomaterial composites, which involves both physical strategies and chemical approaches, such as dry mechanical exfoliation (e.g., Scotch tape), chemical (e.g., solution-based exfoliation, graphite oxide exfoliation/reduction) and/or electrochemical (oxidation/reduction and exfoliation) processes, chemical vapor deposition (CVD), chemical synthesis, thermal decomposition of SiC wafers and unzipping carbon nanotubes [26,30–43]. In this review, we will not describe the detailed synthetic methods/protocols mentioned above for synthesis of 2D nanomaterials, however, we suggest reading several recently published comprehensive review articles [40–45]. These methods of 2D nanomaterial preparation produce different forms of nanomaterials with a diversity of properties including mechanical, optical, electrical, chemical and biological properties. These diverse properties make 2D nanomaterials suitable for an extensive range of applications, such as drug delivery, in vitro and in vivo imaging, tissue engineering, biosensor construction, and energy conversion and storage [39,43,46]. For biosensor applications, 2D nanomaterials should be extensively characterized because their properties strongly dependent on their characteristics such as thickness or number of layers, morphology, chemical structure and surface functional groups.

Microbial toxins are the general term for a class of substances covering a broad range from small molecules to biomacromolecules (e.g., peptides and proteins), which are produced by living organisms including bacteria, fungus and algae [47–53]. They are widespread throughout the whole world, threatening the health and/or life of humans and livestock, and affecting domestic and international trade. For instance, aflatoxin B1 (AFB1, a kind of mycotoxin produced by fungi) has been defined as a group I carcinogen by the World Health Organization (WHO) [53]. Some microbial toxins can generate acute poisonous effects even at very low doses, and the co-occurrence of microbial toxins in nature may cause significantly additive and/or synergistic toxicity. In order to efficiently avoid potential hazards on public health and safety, it is important to precisely and reliably determine the toxins in practical samples from different sources. Liquid chromatography-based methods including high-performance liquid chromatography (HPLC) and high-performance liquid chromatography-tandem mass spectrometry (HPLC/MS/MS) are the gold standards for accurate analysis of toxins [54–59]. Although the HPLC-based methods have high reliability and accuracy, they typically require expensive laboratory facilities and instruments, complex pre-treatment processing of the sample and well-trained operators. These drawbacks strongly limit the application of HPLC-based methods in on-site detections of toxin. Various sensing systems such as surface plasmon resonance (SPR) biosensors, electrochemical biosensors, fluorescence biosensors, colorimetric assays, competitive enzyme-linked immunosorbent assay (ELISAs) and microfluidic immunoassay have been developed for analysis of toxins from different sources including clinical samples, foods, water and feeds [60–66]. Among these biosensing systems, electrochemical biosensors and biotransducers are more attractive because they offer several advantages such as high sensitivity, operational simplicity, relatively low cost, easily miniaturization and suitable on-site analysis [8,11–19,67–69]. These advantages make electrochemical biosensors/transducers of microbial toxins powerful tools in many areas including food, environmental and medical monitoring, disease diagnosis and anti-terrorism security. Owing to the large surface areas and excellent conductivities, the integration of 2D nanomaterials (e.g., graphene and TMDs) and their nanocomposites with electrochemical transducers has great potential to enhance the analytical performance of electrochemical biosensors for detection of toxins [8,11–19]. For example, since its birth multiple research initiatives on graphene applied to electroanalytical chemistry have been launched worldwide, and analysts have been developing a plethora of different graphene-based electrochemical sensing platforms for detection of various targets including microbial toxins. Typically, these electrochemical biosensors comprise a graphene and/or a graphene derivative/nanocomposite-modified electrode as an electrochemical signal transduction element, and a biological recognition element (e.g. antibodies, aptamer and microbial cells). The signal from the biological recognition event is converted to a quantifiable electrical signal because the biological target is normally in close contact with the electrochemical signal transduction element through physical or chemical interactions (e.g., electrostatic interactions, π-π interactions and covalent bonds). Because of their unique properties (e.g., large surface area and good conductivity), the detection performance of an electrochemical biosensor can be significantly improved by using the graphene and/or a graphene derivative/nanocomposite. Therefore the scope of application of 2D nanomaterial-based electrochemical biosensors has been constantly expanding in the field of toxin detection. Some of these studies have been reviewed elsewhere with a focus on the fabrication and toxin detection of graphene-based electrochemical biosensors or as subclassifications in more generalized overviews of the nanomaterial-based electrochemical biosensors [8,11–19]. In this review, we will focus on the recent development of GO/rGO and/or MoS_2/$MoSe_2$-based electrochemical biosensors for the determination of various microbial toxins, such as bacterial toxins, fungal toxins and algal toxins, highlighting some of their current achievements, technical challenges/limitations and the future directions by means of a set of selected recent publications.

2. Detection of Bacterial Toxins

2.1. Botulinum Neurotoxins

The *Botulinum* neurotoxins (BoNTs), which are produced by *Clostridium botulinum*, an anaerobic bacterium, are among the most toxic of all naturally occurring substances [70–72]. Based on their molecular structures, BoNTs are categorized into seven serotypes (from A to G). They inhibit acetylcholine release from presynaptic nerve terminals at the neuro-muscular junction in both the central and peripheral nervous systems through cleavage of soluble N-ethylmaleimide-sensitive factor attachment protein receptors (SNAREs), resulting in flaccid muscle paralysis. BoNTs can cause the deadly disease, botulism, with a median lethal dose (LD50) of 1 ng per kg bodyweight. Foods are easily contaminated by *Clostridium botulinum* during processing. Various (impedimetric, voltammetric and amperometric) electrochemical biosensors have been fabricated for BoNT detection [73–76] In particular, electrochemical biosensors can achieve detection of this toxin in a fast and meticulous way and they also provide a robust and cost-effective approach for real-time monitoring of BoNTs. Recently 2D nanomaterial-based electrochemical biosensors have been applied to sensitively detect BoNTs in various samples including foods. For instance, Narayanan et al. constructed an electrochemical immunosensor of the BoNT serotype E (BoNT/E) by using graphene nanosheets–aryldiazonium salts as transducers [74]. The as-proposed immunosensor shows a low limit of detection (LOD, 5 pg mL^{-1}) and can be employed for rapid detection of BoNT/E with a total analysis time of 65 min Chan et al. fabricated an electrochemical biosensor for ultrasensitive detection of BoNT serotype A light chain (BoNT-LcA) through immobilization of the SNAP-25-GFP (synaptosomal associated protein 25-green fluorescent protein) peptide substrate on the rGO modified gold electrode via a pyrenebutyric acid (PA) linker (as shown in Figure 1) [75]. In this case, PA was immobilized on the rGO surface through π-π stacking. Subsequently, SNAP-25-GFP peptide reacted with PA via N-(3-dimethylaminopropyl)-N′-ethylcarbodiimide hydrochloride/N-hydroxysulfosuccinimide (EDC/Sulfo-NHS) activation. After specific cleavage of SNAP-25-GFP by BoNT-LcA, the steric hindrance and electrostatic repulsion of SNAP-25-GFP decreased, resulting in an increase in the electrochemical signal. The amount of BoNT-LcA can be detected through the change of peak current of the electrochemical redox probe (ferricyanide, $[Fe(CN)_6]^{3-/4-}$(1:1)) by the differential pulse voltammetry (DPV) measurement. The as-fabricated electrochemical biosensor provides a relatively wide linear range (1 pg mL^{-1} to 1 ng mL^{-1}) and a relatively low LOD (8.6 pg mL^{-1}) for detection of BoNT-LcA because the rGO modified Au (rGO/Au) electrode provides a robust and biocompatible platform with improved electron transfer capability and a large surface area for peptide immobilization. The feasibility of the as-fabricated biosensor is demonstrated by detection of BoNT-LcA in spiked milk samples. Afkhami et al. developed a gold nanoparticle-graphene-chitosan (Au NPs-Gr-Cs) nanocomposite-based impedimetric immunosensor for the detection of BoNT serotype A (BoNT/A) [76]. The Au NPs-Gr-Cs nanocomposite was used for the amplification of the electrochemical signal, and monoclonal anti-BoNT/A antibodies were conjugated on the Au NPs-Gr-Cs nanocomposite modified glassy carbon electrode (GCE). In the presence of BoNT/A, the immunocomplex formed on the as-prepared electrode surface, which acts as the inert electron and mass transfer blocking layer. Therefore, the diffusion of $[Fe(CN)_6]^{3-/4-}$ is hindered, resulting in a decrease of the peak current. The Au NPs-Gr-Cs nanocomposite-based impedimetric immunosensor has an excellent linear range (from 0.27 to 268 pg mL^{-1}) with a LOD of 0.11 pg mL^{-1}, and is very suitable for routine analysis of BoNT/A in different matrices, such as serum and milk.

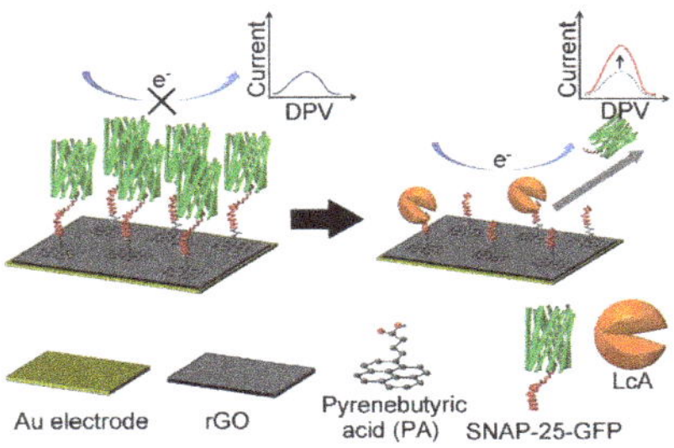

Figure 1. Schematic representation of the detection principle of the rGO based electrochemical biosensors (adapted from Chan et al. 2015 [75], Copyright 2015 Elsevier B.V. and reproduced with permission).

2.2. *Clostridium difficile* Toxin B

Clostridium difficile toxin A (Tcd A, 308 kDa) and toxin B (Tcd B, 270 kDa) are co-produced by *Clostridium difficile (C. difficile)*. Tcd A is an enterotoxin responsible for tissue damage, while Tcd B is referred to as a potent cytotoxin [77–81]. In particular, the rapid and sensitive detection of Tcd B is very helpful for early diagnosis and efficient therapy because Tcd B is critical for virulence and is found in all clinically isolated pathogenic strains [79–85]. Using the advantages of GO, including the large surface area and good conductivity, Fang et al. developed a simple sandwich-assay type electrochemical immunosensor for improving the Tcd B detection sensitivity by using GO as a scaffold for the enhanced loading of horseradish peroxidase (HRP) and HRP-labeled secondary Tcd B antibody (as shown in Figure 2) [84]. The LOD (0.7 pg mL^{-1}) of the sandwich-assay type electrochemical immunosensor is much lower than those of other current techniques including ELISA. In addition, the as-prepared electrochemical immunosensor was successfully employed to detect Tcd B in practical samples (e.g., real human stool), demonstrating that the immunosensor has promising potential in clinical applications.

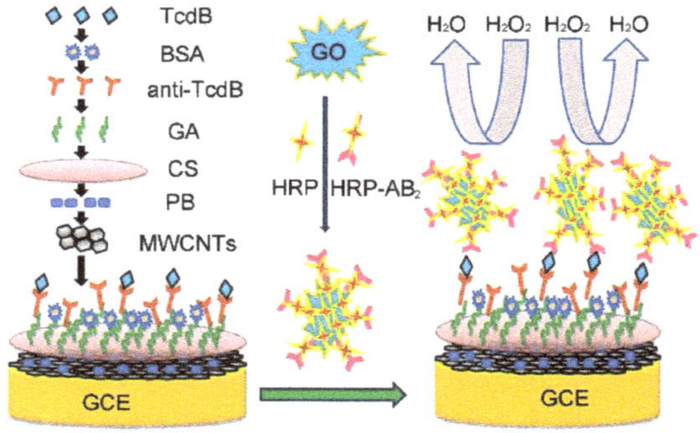

Figure 2. Schematic representation of the immunosensor array preparation and detection strategy by sandwich-type immunoassay of Tcd B. Here, Tcd B means *C. difficile* toxin B, BSA means bovine serum albumin, anti-Tcd B means anti-Tcd B antibody, HRP means horseradish peroxidase, HRP-Ab2 means HRP-labeled second anti-Tcd B antibody, GA means glutaraldehyde, CS means chitosan, PB means Prussian blue, MWCNTs means multi-walled carbon nanotube, GO means graphene oxide, and GCE means glassy carbon electrode (adapted from Fang et al. 2014 [84], Copyright 2013 Elsevier B.V. and reproduced with permission).

2.3. *Staphylococcal Enterotoxin B*

Among the toxins secreted by *Staphylococcus aureus*, the staphylococcal enterotoxin B (SEB) shows superantigenic properties in nature. SEB exposure can result in immunosuppression and serious food poisoning [86,87]. Therefore, it is important to develop a cost-effective, easy-to-use, rapid and sensitive method for real-time monitoring of a low concentration (less than 20 ng kg^{-1} (i.e., LD_{50} value)) of SEB in foods. Several graphene-based electrochemical biosensors have been developed for real-time detection of SEB in foods with a high sensitivity [88–91]. For instance, Sharma et al. reported on an electrochemical biosensor based on a rGO-chitosan-AuNPs-capturing antibody (rGR-Ch-AuNPs-CAb)-modified GCE for detecting SEB [88]. The rGR-Ch-AuNPs-CAb modified GCE shows remarkable detecting performance because it has a flat two-dimensional configuration and large surface area with plenty of active sites (i.e., functional groups). Using the as-proposed rGR-Ch-AuNPs-CAb-based electrochemical biosensor, 5 ng mL^{-1} SEB can be easily detected within 35 min, which is much lower than the LD_{50} value of SEB. Very recently, Nodoushan et al. fabricated an electrochemical aptasensor for SEB detection by using a rGO and gold nano-urchins (AuNUs)-modified screen printed carbon electrode (SPCE) (as shown in Figure 3) [91]. The aptamer of SEB was attached on the electrode surface through hybridization with the immobilized single-stranded DNA probe on the surface of the AuNUs. Hematoxylin was used as the electrochemical signal generator. In the presence of SEB, the aptamer released from the electrode surface, resulting in an increase in the peak current of hematoxylin. Benefiting from the high conductivity of rGO and high surface area of AuNUs, a wide linear range from 5.0 to 500.0 fmol L^{-1} was achieved and the LOD was calculated as 0.21 fmol L^{-1}. There is no significant difference between the results given by the commercial ELISA kit and the electrochemical aptasensor. In particular, the aptasensor shows better recovery rates and lower standard deviation than those of the commercial ELISA kit, which could be employed as a point-of-care (POC) device for assessing food samples.

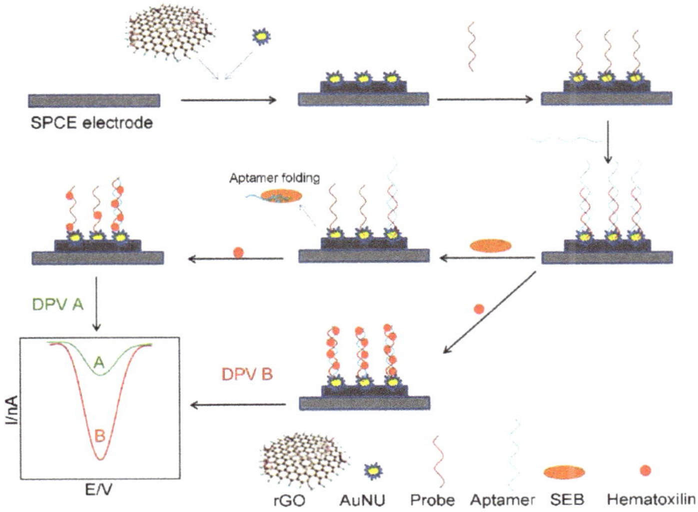

Figure 3. Schematic representation of the fabrication process of the SEB aptasensor by using rGO and AuNU-modified screen printed carbon electrodes (SPCEs) (adapted from Nodoushan et al. 2019 [91], Copyright 2018 Elsevier B.V. and reproduced with permission).

3. Detection of Fungal Toxins

3.1. *Aflatoxins*

Aflatoxins are a widespread group of food toxins that are produced by *Aspergillus flavus* and *Aspergillus parasiticus* [92–95]. There are four main types of aflatoxins: B_1, B_2, G_1, and G_2, which are based on their fluorescence characteristics under UV light (blue or green) excitation and relative chromatographic mobility in thin-layer chromatography. Among the aflatoxins, AFB1 is considered the most toxic aflatoxin, and can cause cancers, such as hepatocellular carcinoma. Various 2D nanomaterial-based electrochemical biosensors have been constructed for detecting AFB_1 in various matrixes [96–113]. Srivastava et al. have developed a series of functionalized GO nanocomposite-based electrochemical biosensors for profiling AFB_1 in foods since they developed the first rGO-based AFB_1 immunosensor through the covalent conjugation of the monoclonal anti-AFB1 antibodies onto an rGO modified indium tin oxide (ITO) electrode in 2013 [96–99]. Among these electrochemical biosensors, the functionalized GO/rGO-based nanocomposites are employed in different roles, such as catalysts, electroactive probes and immobilization platforms for improving the biosensing performance. For instance, benefiting from the highly crystalline properties of the rGO-Ni NPs sheets (Ni nanoparticle decorated rGO sheets) along with the excellent electro-catalytic properties, the rGO-Ni NPs-ITO-based AFB_1 immunosensor exhibits high sensitivity (129.6 mA ng^{-1} mL cm^{-2}), long term stability (up to 6 weeks) and low LOD (0.16 ng mL^{-1}) [99]. Photoelectrochemical (PEC) biosensors have attracted great attention in the biological analytical field as the PEC method can obtain high sensitivity without expensive equipment. Recently, Hao et al. developed a dual channel self-reference PEC biosensor for detecting AFB_1 through immobilization of the AFB_1 aptamer onto cadmium telluride (CdTe) and the CdTe-GO modified ITO electrode (as shown in Figure 4) [104]. In this case, CdTe and CdTe-GO were used to generate an anodic photocurrent and cathodic photocurrent, respectively. The AFB1 aptamer was immobilized on the PEC active materials, CdTe and CdTe-GO, through a covalent reaction or physical absorption, respectively. In the presence of AFB1, the aptamer is released from the CdTe-GO surface, resulting in the recovery of the cathodic photocurrent, while the aptamer forms an aptamer-AFB1 complex on the CdTe surface, and the anodic photocurrent decreases further. Compared to traditional PEC biosensors, the CdTe/CdTe-GO-based dual channel self-reference PEC biosensor can provide better precision and reliability, which is promising for detection of AFB_1 in complex matrixes. Very recently, Peng et al. developed an AFB1 electrochemical aptasensor based on tetrahedral DNA nanostructures (TDNs) immobilized on three dimensionally ordered macroporous MoS_2-AuNPs hybrids (3DOM MoS_2-AuNPs) [107]. 3DOM MoS_2-AuNPs can enhance the immobilization amount of TDNs and facilitate the movement of the electrons between the electrode surface and the redox probe. In combination with a HRP functionalized magnetic signal amplifier, the aptasensor achieves a good linear range (from 0.1 fg mL^{-1} to 0.1 μg mL^{-1}) and a LOD of 0.01 fg mL^{-1}, which can be employed to detect AFB_1 in grain products such as rice and wheat powder samples.

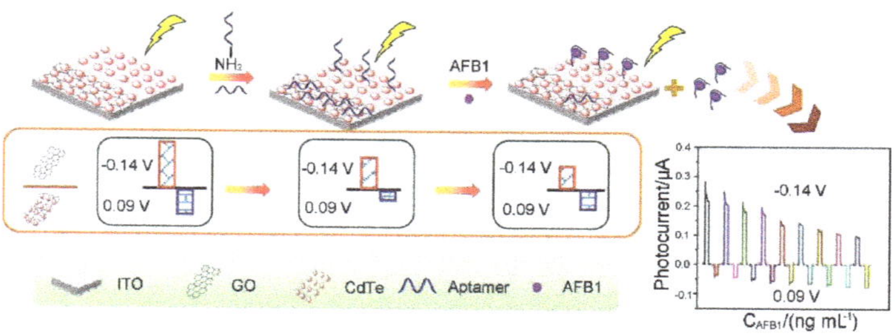

Figure 4. Schematic representation of the construction of the self-reference photoelectrochemical (PEC) biosensor for the detection of AFB_1 (adapted from Hao et al. 2017 [104], Copyright 2017 American Chemical Society and reproduced with permission).

3.2. Ochratoxin

Ochratoxin A (OTA) is the major mycotoxin of the ochratoxin group, which are produced primarily by fungi (e.g., *Aspergillus ochraceus, Penicillium verrucosum* and *Aspergillus niger*) [114–117]. OTA has strong nephrotoxicity, and is the main etiological agent responsible for human Balkan endemic nephropathy (BEN) and associated urinary tract tumors. In addition, high concentrations of OTA has certain hepatotoxicity. During the last 5 years, several 2D nanomaterial-based electrochemical biosensors including immunosensors and aptasensors have also been developed for sensing OTA [118–131]. For instance, a series of aptasensors based on rGO-AuNP nanocomposites have been constructed by Wang's group [118–120]. The rGO-AuNP nanocomposites have well-dispersity and controllable surface coverage of AuNPs on the rGO sheet, which can be employed as an excellent signal amplified platform for an impedimetric aptasensor and/or an efficient nanocarrier for the CdTe QD (cadmium telluride quantum dot)-based amperometric aptasensor. As a typical example, a label free electrochemical aptasensor was successfully fabricated for ultrasensitive detection of OTA through using the CdTe QDs modified graphene/AuNPs nanocomposite (GAu/CdTe) as a signal amplifier. The as-proposed label-free amperometric aptasensor exhibits a wide linear range from 0.2 pg mL^{-1} to 4 ng mL^{-1} and a low LOD (0.07 pg mL^{-1}), which has great potential in various applications, such as food safety monitoring and clinical diagnosis [120]. Bulbul et al. developed a non-enzymatic nanocatalyst-based amperometric aptasensor for OTA detection through immobilization of the OTA aptamer on the GO-modified electrode and the electro-oxidation of a nanoceria (nCe) tag [121]. In this case, GO was used as an electrode material for facilitating the electron transport and enhancing the electrochemical response because it has high conductivity and peroxidase-like activity. In particular, the synergistic effect between the catalase activity of nCe and the peroxidase like activity of GO increases the OTA detection sensitivity significantly. The LOD of as-proposed amperometric aptasensor is calculated to be 0.1 nmol L^{-1}, which is below the European Union regulatory limits of OTA (such as 5 μg kg^{-1} in raw cereal grains, 3 μg kg^{-1} in products derived from cereals, and 2 μg kg^{-1} in grape juice). The analytical reliability of the amperometric aptasensor has been demonstrated by the detection of OTA in spiked corn samples. Recently, Wang et al. constructed a ratiometric electrochemical aptasensor for OTA detection through assembly of a methylene blue (MB)-modified OTA aptamer (MB-aptamer) on the MoS_2 nanosheet/AuNP (MoS_2-AuNP) nanocomposite-decorated gold electrode through the host-guest recognition of β-cyclodextrin (β-CD) (as shown in Figure 5) [128]. After interaction with OTA, the MB-aptamer was disassembled because of G-quadruplex formation, leading to a decrease in the peak current of MB. Whereas the free ferrocenecarboxylic acid was recognized by β-CD and produced signals in the current, resulting in the "ratiometric" effect. With the combination of high electrocatalytic activity of MoS_2-AuNP nanocomposites and the recognition capability of β-CD, the as-proposed ratiometric electrochemical aptasensor possesses satisfactory superiority in terms of detection range (from 0.1 nmol L^{-1} to 50 nmol L^{-1}), sensitivity (a LOD of 0.06 nmol L^{-1}), and accuracy (6.5% of the relative standard deviation (RSD)). The practicability of the aptasensor was successfully demonstrated by detecting OTA in red wine samples.

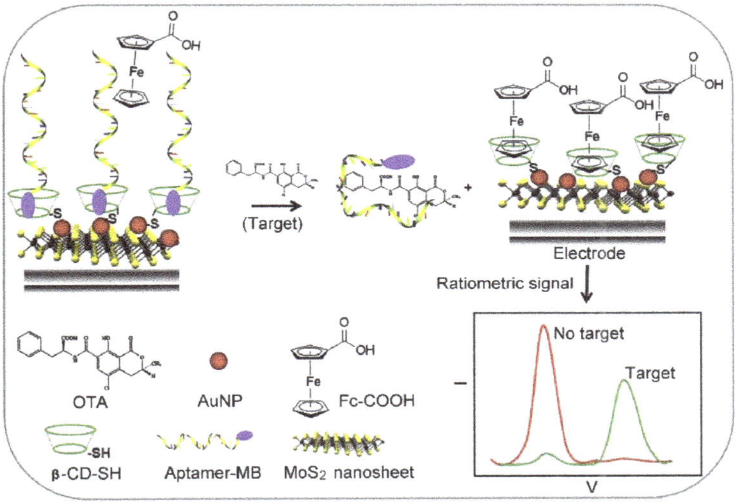

Figure 5. Schematic representation of the fabrication of the ratiometric electrochemical aptasensor for OTA detection based on nanocomposites of gold nanoparticle and MoS_2 nanosheets with β-CD-SH (thiolated β-CD) (adapted from Wang et al. 2018 [128], Copyright 2018 Elsevier Ltd. and reproduced with permission).

3.3. *Mycotoxins Produced by Fusarium*

The 2D nanomaterial-based electrochemical biosensors have also been developed for detecting other *mycotoxins* produced by *Fusarium* including deoxynivalenol (DON), fumonisin 1 (FB1), and zearalenone (ZEN) [132–137]. Shi et al. developed an aptasensor for sensitive FB1 detection by using the dual amplification of AuNPs and graphene/thionine nanocomposites (GSTH) [132]. GSTH served as electrochemical probes, which exhibit a strong electrochemical signal because the graphene has excellent conductivity and a large surface area for immobilizing a large amount of thionine molecules. The as-prepared aptasensor has a six orders of magnitude linear range with a LOD of 1 pg mL^{-1}. Lu et al. fabricated an electrochemical immunosensor based on a graphene nanocomposite for rapid and sensitive detection of two mycotoxins, DON and FB1 by using correspondent anti-toxin antibodies (as shown in Figure 6) [134]. In this case, the disposable SPCE was used as a sensing platform, which was modified by AuNPs and polypyrrole (PPy)-electrochemical rGO (PPy/ErGO) nanocomposite film. The film exhibits effective anti-toxin antibody immobilization capacity, enhanced electrical conductivity, and biocompatibility. The current signal of PPy/ErGO-SPCE is much better than that of PPy/rGO-SPCE. Benefiting from the excellent electrochemical response and effective antibody immobilization, the immunosensor exhibits good sensitivity, with a LOD of 4.2 ng mL^{-1} for FB1 and 8.6 ng mL^{-1} for DON. The immunosensor can be used for simultaneous detection of multiple co-contaminant mycotoxins individually in the practical samples (e.g., corn extracts) because it shows low matrix interference even in co-existing toxin environments. Very recently, Jiang et al. constructed a facile electrochemical immunosensor based on thin-layer MoS_2 and thionin (MoS_2-Thi) composites for the sensitive and rapid detection of zearalenone (ZEA) in human biofluids (as shown in Figure 7) [136]. The as-prepared MoS_2-Thi nanocomposites were employed as excellent electrochemical probes, as well as an efficient anti-ZEA antibody loading platform because MoS_2 retains the electrochemical activity of Thi, and has a large surface area. The MoS_2-Thi-based electrochemical immunosensor has good ZEA detection performance including a wide linear range (0.01 to 50 ng mL^{-1}), low LOD (0.005 ng mL^{-1} ZEA in both the plasma and urine), excellent selectivity, rapid responding time (20 min), acceptable stability (retained more than 85% detection capability at 4 °C for 10 days) and good practicability (detection of ZEA in real human biofluids).

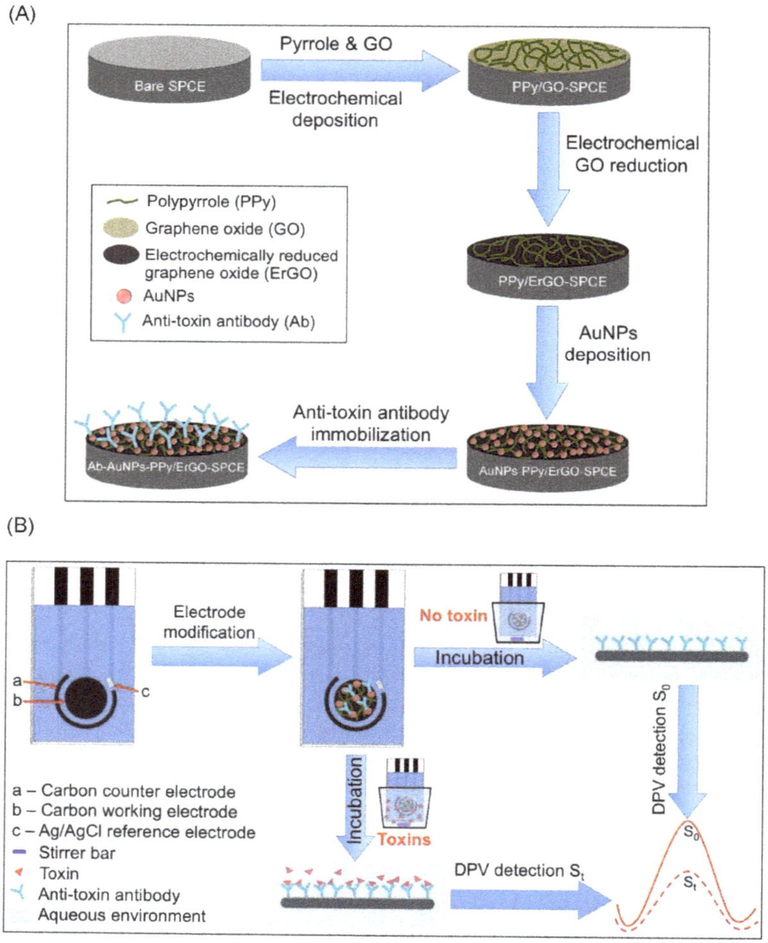

Figure 6. (**A**) Schematic representation of fabrication of the immunosensor and (**B**) detection of mycotoxins (adapted from Lu et al. 2016 [134], Copyright 2016 Elsevier Ltd. and reproduced with permission).

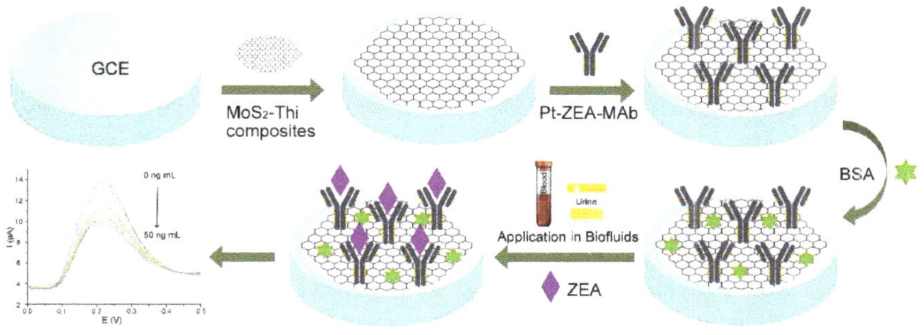

Figure 7. Schematic representation of the electrochemical immunosensor based on MoS_2-Thi composites for the rapid detection of ZEA in biofluids (adapted from Jiang et al. 2019 [136], Copyright 2019 Elsevier B.V. and reproduced with permission).

4. Detection of Algal Toxins

4.1. Microcystins

Microcystins (MCs), a group of toxins produced by a number of cyanobacteria species, are monocyclic heptapeptides with the general structure cyclo(D)-Ala-X-(D)-erythro-b-methyl-iso-Asp-Y-Adda-(D)-iso-Glu-N-meth-yldehydro-Ala (X and Y represent L-amino acids). They are the most common cyano-toxins [138–142]. The unusual Adda amino acid, unique to MCs, is responsible for the toxicity of the molecule. There are more than 100 known variants of MCs, which are found in a wide variety of aquatic environments, in particular, eutrophic waters. Exposure to MCs via consumption of poisoned drinking-water or eating contaminated fish can cause permanent multiple organ injuries, developmental effects, reproductive effects and cancer. Therefore, it is important to develop highly sensitive methods for on-site monitoring of MCs. In addition, as the most potent congener, the Microcystin-LR (MC-LR) is commonly used to evaluate the toxicological data on the effects of MCs. The maximum tolerance limit of MC-LR concentration is 1 $\mu g\ L^{-1}$ in different water sources by the WHO provisional guideline. Electrochemical biosensors, including 2D nanomaterial-based amperometric immunosensors, impedimetric aptasensors, and PEC aptasensors, have been extensively employed to detect MCs/MC-LR [143–159]. Li et al. have fabricated an electrochemical immunosensor based on GO-AuNP nanocomposites for MC-LR detection in water samples though layer-by-layer alternate electrodeposition of GO and chloroauric acid ($HAuCl_4$) on the GCE surface for 20 cycles [147]. The GO-AuNP-decorated GCE was then modified by the conducting polymer (poly(2,5-di-(2-thienyl)-1-pyrrole-1-(p-benzoicacid)) and 1-iso-butyl-3-methylimidazolium bis(tri-fluoromethane-sulfonyl) imide ionic liquid (IL). A polyclonal antibody of MC-LR was immobilized on the electrode by the conventional EDC/NHS reaction. The GO-AuNP nanocomposites enhance electron transfer of $Fe(CN)_6{}^{3-/4-}$ to the electrode while the IL acts as the stabilizer of the antibody. The as-developed electrochemical immunosensor has good repeatability (e.g., RSD = 1.2%) and long-term stability (e.g., retain 95% activity over a 20 weeks storage period), and can detect MC-LR in water samples with a very low LOD of 3.7×10^{-17} mol L^{-1}. Recently, He et al. synthesized a kind of magnetic rGO nanocomposite (Fe_3O_4@PDA/RGO) for constructing a MC-LR electrochemical immunosensor by using the hydrothermal treatment of Fe_3O_4 nanocluster@Polydopamine core@shell nanoparticles (Fe_3O_4@PDA) with GO (as shown in Figure 8) [153]. Due to its surface area and easy separation, the Fe_3O_4@PDA/RGO clearly enhances the antigen immobilization ability of the electrode. Then, a secondary-antibody and circularization DNA template were conjugated on gold nanorods (AuNRs) for recognizing the captured MC-LR-antibody pair on the Fe_3O_4@PDA/RGO-modified electrode surface and rolling circle amplification. Because the rolling circle amplification strategy can generate massive repeated DNA sequences, the signal of the immunosensor is greatly enhanced by hybridization of electrochemical active probes with the repeated DNA sequences. Under the optimal conditions, the as-developed immunosensor has good detection performance including a wide linear range (from 0.01 mg L^{-1} to 50 mg L^{-1}) and a low LOD (0.007 mg L^{-1}), which can be employed to detect MC-LR in real samples (e.g., river water). A series of PEC aptasensor-based various GO/rGO nanocomposites have been developed for sensitively detecting MC-LR since the PEC method has been considered to be a more sensitive technique, ascribed to the combination of electrochemical and optical techniques [149,151,157]. For instance, Du et al. developed a PEC aptasensing platform based on AgI-nitrogen-doped graphene (AgI-NG) nanocomposites as photo-cathodes and a MC-LR aptamer as the recognition unit [157]. The PEC aptasensor has a LOD of 3.7×10^{-17} mol L^{-1}, which can be employed to determine MC-LR in inaquatic products (e.g., fish extracts). As a graphene analogue, the MoS_2 nanosheet is also expected to serve as an excellent functional material for development of electrochemical biosensors. As shown in Figure 9, Pang et al. constructed an enzyme-free electrochemical immunosensor for detecting MC-LR based on a unique competitive detection scheme using MoS_2 nanosheets/BSA-stabilized gold nanocluster (MoS_2/AuNCs) nanocomposites and Au core/Pt shell nanoparticles (Au@PtNPs) [155]. Due to its large surface area and excellent biocompatibility, the MoS_2/AuNCs nanocomposite was employed as a platform for improving the biological activity and immobilizing amount of antibody on the electrode surface. The as-developed enzyme-free electrochemical immunosensor has good stability (e.g., 92% of the initial level remained after being stored at 4 °C for four weeks), and exhibits a wide linear range of 1.0 ng L^{-1}–1.0 mg L^{-1} with a LOD of 0.3 ng L^{-1}. The practicability of the as-developed immunosensor has been demonstrated by detection of MC-LR in various water samples including tap water, lake water, and river water. The MC-LR amounts in these water samples detected by the immunosensor are consistent with those determined by the conventional ELISA method. Very recently, Liu et al. developed an electrochemical aptasensor for sensitive and selective determination of microcystin-LR by using a dual signal amplification system consisting of a ternary nanocomposite and HRP [159]. The ternary nanocomposites were prepared by depositing AuNPs on the MoS_2 nanosheets covered with TiO_2 nanobeads (TiONBs). The MoS_2 nanosheet-modified TiONBs provide a large surface area for efficiently immobilizing AuNPs and thiolated MC-LR aptamers. Due to the combination of good electron transfer and high catalytic capability of the ternary composite, the aptasensor has a wide dynamic range from 0.005 to 30 nmol L^{-1} and a LOD of 0.002 nmol L^{-1}.

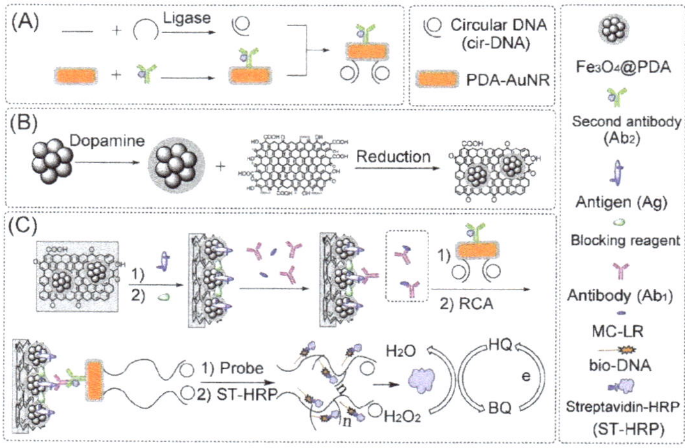

Figure 8. Schematic representation of (**A**) the preparation of Ab_2-AuNR-cirDNA, (**B**) the formation of magnetic graphene composite, and (**C**) the construction process of the proposed MC-LR immunosensor (adapted from He et al. 2017 [153], Copyright 2017 The Royal Society of Chemistry and reproduced with permission).

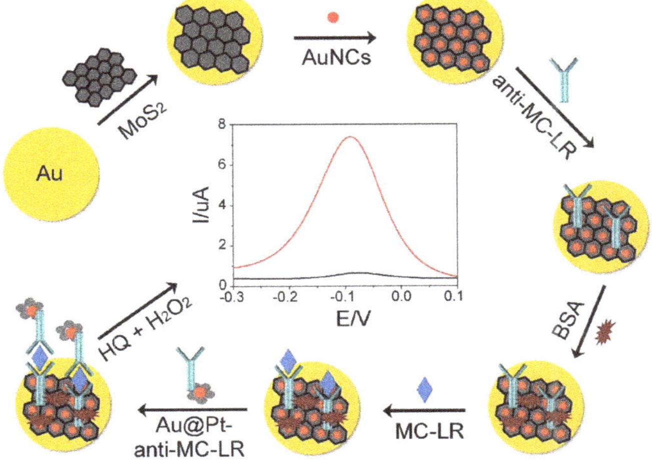

Figure 9. Schematic representation of the preparation and detection principle of the MC-LR immunosensor (adapted from Pang et al. 2018 [155], Copyright 2018 Elsevier B.V. and reproduced with permission).

4.2. *Cylindrospermopsin*

Cylindrospermopsin (CYN), a tricyclic alkaloid with a molecular mass of 415 Da, is a common cyanotoxin, and is produced by cyanobacteria including *Cylindrospermopsis*, *Anabaena*, *Umezakia*, and *Aphanizomenon* [160–166]. Cylindrospermopsin can cause DNA/RNA strand breakage and promote hepatotoxicity, cytoxicity, and genotoxicity through inhibiting protein translation and binding to DNA. The Falconer recommends a tentative guideline value of 1 ug L^{-1} for cylindrospermopsin [166]. Recently, we fabricated a label-free impedimetric aptasensor based on a GO-thionine (TH-GO) nanocomposite for detection of CYN by covalent binding of the amino-terminated aptamer of CYN to TH-GO nanocomposite-modified GCE via glutaraldehyde (as shown in Figure 10) [167]. Using $[Fe(CN)_6]^{4-/3-}$ as an electrochemically active probe, CYN can be detected as low as 0.117 ng mL^{-1} in water. The as-proposed aptasensor has been employed for detecting CYN in spiked lake water samples, and satisfactory recoveries were obtained. With its superior performance characteristics combined with long-term stability (it retained approximately 74.7% of its initial value after being stored at 4 °C for 30 days) and excellent reusability (RSD = 2.1% within 10 reacting cycles), the as-developed aptasensor is a potential candidate for on-site CYN analysis.

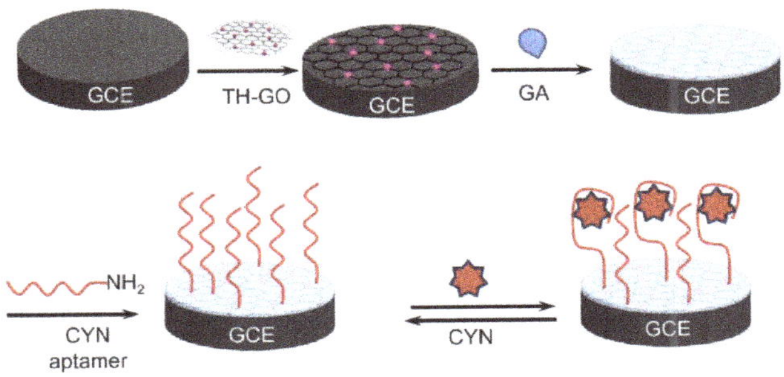

Figure 10. Schematic representation of the label-free impedimetric aptasensor for detecting cylindrospermopsin (adapted from Zhao et al. 2015 [167], Copyright 2015 The Royal Society of Chemistry and reproduced with permission).

4.3. Saxitoxins

As a group of carbamate alkaloid neurotoxins, saxitoxins (STXs) contain sixteen variants, which are commonly associated with "red tides," and found as a paralytic shellfish toxin. Australia has a drinking water guideline of 3 μg L^{-1} of STX equivalence. Recently, Bratakou et al. constructed a miniaturized potentiometric STX immunosensor on graphene nanosheets with incorporated lipid films and anti-STX (the natural STX receptor) [168]. The potentiometric STX immunosensor can be easily miniaturized because graphene nanosheets have a high surface area and good conductivity, and incorporate well with the lipid bilayer membrane for immobilizing anti-STX antibody. The potentiometric STX immunosensor exhibits several advantages such as a rapid response time (ca. 5–20 min), low LOD (1 nmol L^{-1}) with high sensitivity (ca. 60 mV/decade of toxin concentration), good reproducibility (maximum deviation only 6.8%), reusability, high selectivity and long shelf life (> 1 month). The practicability of the method was demonstrated by detecting STX in lake water and shellfish samples. This graphene nanosheets with incorporated lipid films could be used to develop biosensors for monitoring other toxins.

4.4. Brevetoxin B

Brevetoxins (BTXs) are potent cyclic polyether neurotoxins, which are naturally produced by the marine "red tide" dinoflagellate, *Karenia brevis*. BTX exposure can cause neurological shellfish poisoning (NSP), which has increased in geographical distribution over the past decade [139]. As early as 2012, Tang et al. constructed a magneto-controlled electrochemical immunosensor for sensitive detection of brevetoxin B (BTX-2) in seafood by using guanine-assembled graphene nanoribbons (GGNRs) as molecular tags on a home-made magnetic carbon paste electrode [169]. In this case, the GGNRs were modified by bioconjugates of BSA with BTX-2 (BTX-2-BSA), while monoclonal mouse anti-BTX-2 antibodies were covalently immobilized on the surface of magnetic beads for the capture of BTX-2 through a competitive-type immunoassay format. The formed magnetic immunocomplex was integrated on the electrode with an external magnet, followed by determination in pH 6.5 phosphate-buffered solution containing 2 μmol L^{-1} $Ru(bpy)_3Cl_2$. Compared with pure guanine-labeled molecular tags, the GNR-labeled electrochemical immunoassays show a much wider linear range and lower detection limit. Under optimal conditions, the electrochemical signals decreased by increasing concentration of BTX-2 in the sample. The magneto-controlled immunosensing platform has a wide dynamic range from 1.0 pg mL^{-1} to 10 ng mL^{-1} with a LOD of 1.0 pg mL^{-1} BTX-2. The analytical reliability of the magneto-controlled electrochemical immunosensing platform is demonstrated by the detection of BTX-2 in 12 spiked samples including *S. constricta, M. senhousia and T. granosa*. The as-obtained results are consistent with those of traditional ELISA.

4.5. Okadaic Acid

The family of okadaic acid (OA) biotoxins consists of OA and its analogues dinophysistoxins 1, 2 and 3 (named as DTX-1, DTX-2 and DTX-3) [170]. As a by-product of harmful algal blooms (HABs), OA originates from the algal genera *Prorocentrum* and *Dynophysis*. Eissa and Zourob developed a direct competitive voltammetric immunosensor for the sensitive detection of OA based on carboxyphenyl-functionalized graphene-modified SPCEs (GSPCEs) [171]. The anti-OA antibodies were immobilized on the GSPE via carbodiimide chemistry, where OA and OA-ovalbumin (OA-OVA in solution compete for their binding to the immobilized antibody. Benefitting from the unique electrochemical properties of graphene and the stability of the carboxyphenyl layer, the immunosensor exhibits a linear response up to 5000 ng L^{-1} with a LOD of 19 pg mL^{-1}. The immunosensor was successfully applied for detecting OA in the spiked shellfish extracts, showing good recovery. Very recently, Ramalingam et al. fabricated an electrochemical microfluidic biochip for detecting OA by using phosphorene-gold (BP-Au) nanocomposite-modified SPCE (as shown in Figure 11) [172]. The as-synthesized BP-Au nanocomposite not only serves as a backbone to the aptamer sequence but also significantly enhances the electrochemical response of the aptasensor. DPV measurements revealed a LOD of 8 pmol L^{-1}, while a linear range was found between 10 nmol L^{-1} to 250 nmol L^{-1}. The electrochemical aptasensor has excellent selectivity and can be employed to detect OA in fresh mussel extracts. The results suggest that the microfluidic electrochemical aptasensor can be served as an easy-to-use POC device for an on-field assay.

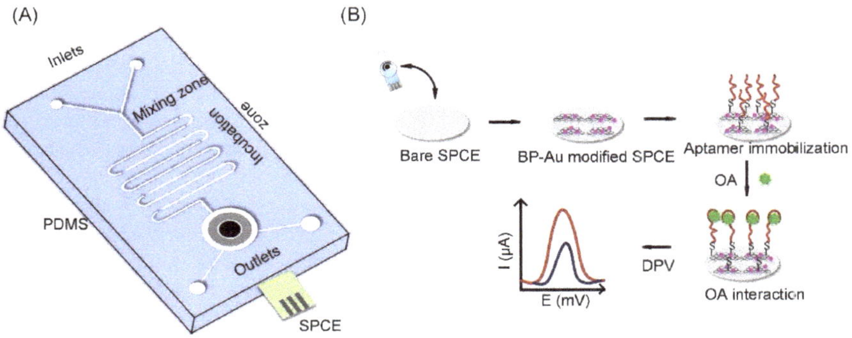

Figure 11. A microfluidic electrochemical aptasensor for the detection of okadaic acid: (**A**) graphic of the fabricated PDMS microfluidic chip, and (**B**) schematic representation of the process of aptamer-based sensing (adapted from Ramalingam et al. 2019 [172], Copyright 2019 Elsevier B.V. and reproduced with permission).

5. Conclusions and Perspective

This review has summarized the recent progress in electrochemical biosensing systems for the determination of various microbial toxins by using 2D nanomaterials and their nanocomposites (hereinafter referred to 2D nanomaterials). The literature results demonstrate that the integration of 2D nanomaterials into electrochemical biosensors has led to the significant enhancement of their analytical efficiency, including a high sensitivity (e.g., very low LODs) with a wide linearity range over several orders of magnitude, rapid assaying time, and simplified analytical procedures, and they are also suitable for on-site monitoring. During the determination processes, 2D nanomaterials mainly have two roles: as substrates for efficient immobilization of capturing biomolecules (e.g., anti-toxin antibodies and aptamers) and high active electrochemical probes for signal amplification. Some 2D nanomaterials have multifunctionality, and are capable of playing both of the above roles. Furthermore, the 2D nanomaterial-based electrochemical aptasensors have been proven as reusable platforms for detecting toxins.

Although the 2D nanomaterial-based electrochemical biosensors show great promise within laboratory investigations, such as the detection of toxins in buffer solutions and/or toxin-spiked samples, the technique remains relatively immature in development compared with standard toxin assaying tools (e.g., HPLC and ELISA), and several technical challenges are still awaiting further investigation. (1) The multiple electrode modification steps are normally required for increasing the recognition performance of the immobilized aptamer or antibody, and reducing background signals. This phenomenon requires manual and tedious work, which not only increases the preparation cost of biosensors, but also leads to poor reproducibility of the results among laboratories. In order to simplify the biosensor construction procedure, future research should increase the reaction efficiency of 2D nanomaterials with biomolecules (such as an antibody and apatmer) and decrease unreacted activity groups on the surface of 2D nanomaterials after biomolecule immobilization. Furthermore, development of automatic methods for modification of 2D nanomaterials on the electrode surface may help to increase the inter-laboratory reproducibility of biosensors. (2) The properties of 2D nanomaterials, including their electrical conductivity, PEC conversion capability and biomolecule immobilization capacity, are strongly dependent on their morphology, such as shape, size, purity, and defects. Therefore, 2D-nanomaterials should be fully characterized before biosensor fabrication. In further research, researchers are strongly encouraged to establish the synthesis standard of 2D-nanomaterials in order to improve the reproducibility of 2D nanomaterial-based electrochemical biosensors. In addition, the as-proposed synthesis strategy should be easily employed to produce 2D-nanomaterials on a large-scale by simply adjusting the synthesis conditions, such as increasing the amount of reactants. This factor is very important for industrialization of the 2D nanomaterial-based electrochemical sensors. (3) To date, one kind of 2D nanomaterial-based electrochemical biosensor is merely confined to determine a single microbial toxin. Because of coexistence of various microbial toxins in nature, future research should focus on development of a universal biosensor production technology for enabling rapid analysis of various toxins. (4) In order to achieve large-scale application, in particular for on-site monitoring, further efforts should be directed toward the development of 2D nanomaterial-based electrochemical biosensors, which can be used to detect toxins in practical samples such as various agricultural, food stuff, body fluids, and environmental sectors (e.g., lake water and sea water). The practicability of 2D nanomaterial-based electrochemical biosensors could be improved through integration of the biosensor with other techniques such as microfluidic devices and microarrays because miniaturization will help to increase the detection throughput, e.g., recognize multiple elements simultaneously. (5) Currently, aptamers and antibodies are mainly used for recognition of the toxins. In order to obtain high selectivity, the key epitope residues of the aptamer and antibody should be unrestrained after immobilization on the 2D nanomaterials. In addition, the molecular structures of the aptamer and antibody are sensitive to the environmental conditions (such as temperature, ionic strength and interferences from sample matrices). The high apparent affinity of the aptamer and/or antibody could be achieved through immobilization of the aptamer and/or antibody by stereoselective reactions (e.g., chick chemistry, DNA hybridization, biotin-avidin recognition). In addition, future research should aim to increase the biocompatibility of 2D nanomaterials. Finally, we expect commercialization of 2D nanomaterial-based electrochemical biosensors into practical procedures for detecting multiple toxins in practical samples through efforts of researchers in different disciplines, which would give significant benefit to the public.

Author Contributions: Conceptualization, Z.L. and X.L. contributed equally for preparation of the draft of this article, M.J. prepared the figures, G.S.G. and Z.W. contributed equally to its further elaboration and discussion. All authors have read and agreed to the published version of the manuscript.

Funding: This research was funded by the National Natural Science Foundation of China (Grant no. 21775145) to Z.W.

Conflicts of Interest: The authors declare no conflict of interest.

References

1. Shao, Y.; Wang, J.; Wu, H.; Liu, J.; Aksay, I.A.; Lin, Y. Graphene based electrochemical sensors and biosensors: A review. *Electroanalysis* **2010**, *22*, 1027–1036. [CrossRef]
2. Brownson, D.A.C.; Banks, C.E. Graphene electrochemistry: An overview of potential applications. *Analyst* **2010**, *135*, 2768–2778. [CrossRef] [PubMed]
3. Yang, W.; Ratinac, K.R.; Ringer, S.P.; Thordarson, P.; Gooding, J.J.; Braet, F. Carbon nanomaterials in biosensors: Should you use nanotubes or graphene? *Angew. Chem. Int. Ed.* **2010**, *49*, 2114–2138. [CrossRef] [PubMed]
4. Kuila, T.; Bose, S.; Khanra, P.; Mishra, A.K.; Kim, N.H.; Lee, J.H. Recent advances in graphene-based biosensors. *Biosens. Bioelectron.* **2011**, *26*, 4637–4648. [CrossRef] [PubMed]
5. Liu, Y.; Dong, X.; Chen, P. Biological and chemical sensors based on graphene materials. *Chem. Soc. Rev.* **2012**, *41*, 2283–2307. [CrossRef] [PubMed]
6. Yang, T.; Yang, R.; Chen, H.; Nan, F.; Ge, T.; Jiao, K. Electrocatalytic activity of molybdenum disulfide nanosheets enhanced by self-doped polyaniline for highly sensitive and synergistic determination of adenine and guanine. *ACS Appl. Mater. Interfaces* **2015**, *7*, 2867–2872. [CrossRef]
7. Kannan, P.K.; Late, D.J.; Morgan, H.; Rout, C.S. Recent developments in 2D layered inorganic nanomaterials for sensing. *Nanoscale* **2015**, *7*, 13293–13312. [CrossRef]

8. Reverte, L.; Prieto-Simon, B.; Campas, M. New advances in electrochemical biosensors for the detection of toxins: Nanomaterials, magnetic beads and microfluidics systems. A review. *Anal. Chim. Acta* **2016**, *908*, 8–21. [CrossRef]
9. Kukkar, M.; Tuteja, S.K.; Sharma, A.L.; Kumar, V.; Paul, A.K.; Kim, K.H.; Sabherwal, P.; Deep. A. A new electrolytic synthesis method for few-layered MoS_2 nanosheets and their robust biointerfacing with reduced antibodies. *ACS Appl. Mater. Interfaces* **2016**, *8*, 16555–16563. [CrossRef]
10. Lim, S.A.; Ahmed, M.U. Electrochemical immunosensors and their recent nanomaterial-based signal amplification strategies: A review. *RSC Adv.* **2016**, *6*, 24995–25014. [CrossRef]
11. Hu, Y.; Huang, Y.; Tan, C.; Xiao, Z.; Lu, Q. Two-dimensional transition metal dichalcogenide nanomaterials for biosensing applications. *Mater. Chem. Front.* **2016**, *1*, 24–36. [CrossRef]
12. Justino, C.I.L.; Comes, A.R.; Freitas, A.C.; Duarte, A.C.; Rocha-Santos, T.A.P. Graphene based sensors and biosensors. *TrAC Trends Anal. Chem.* **2017**, *91*, 53–66. [CrossRef]
13. Wang, Y.H.; Huang, K.J.; Wu, X. Recent advances in transition-metal dichalcogenides based electrochemical biosensors: A review. *Biosens. Bioelectron.* **2017**, *97*, 305–316. [CrossRef] [PubMed]
14. Gan, X.; Zhao, H.; Quan, X. Two-dimensional MoS_2: A promising building block for biosensors. *Biosens. Bioelectron.* **2017**, *89*, 56–71. [CrossRef] [PubMed]
15. Goud, K.Y.; Kalisa, S.K.; Kumar, V.; Tsang, Y.F.; Lee, S.E.; Gobi, K.V.; Kim, K.H. Progress on nanostructured electrochemical sensors and their recognition elements for detection of mycotoxins: A review. *Biosens. Bioelectron.* **2018**, *121*, 205–222. [CrossRef] [PubMed]
16. Jiang, C.; Lan, L.; Yao, Y.; Zhao, F.; Ping, J. Recent progress in application of nanomaterial-enabled biosensors for ochratoxin A detection. *TrAC Trends Anal. Chem.* **2018**, *102*, 236–249. [CrossRef]
17. Lv, M.; Liu, Y.; Geng, J.; Kou, X.; Xin, Z.; Yang, D. Engineering nanomaterials-based biosensors for food safety detection. *Biosens. Bioelectron.* **2018**, *106*, 122–128. [CrossRef]
18. Jin, W.; Maduraiveeran, G. Nanomaterial-based environmental sensing platforms using state-of-the-art electroanalytical strategies. *J. Anal. Sci. Technol.* **2018**, *9*, 18. [CrossRef]
19. Chen, Q.; Zhu, L.; Chen, J.; Jiang, T.; Ye, H.; Ji, H.; Tsang, S.; Zhao, Z.; Yi, T.; Chen, H. Recent progress in nanomaterial-based assay for the detection of phytotoxins in foods. *Food Chem.* **2019**, *277*, 162–178. [CrossRef]
20. Krishnan, S.K.; Singh, E.; Singh, P.; Meyyappan, M.; Nalwa, H.S. A review on graphene-based nanocomposites for electrochemical and fluorescent biosensors. *RSC Adv.* **2019**, *9*, 8778–8881. [CrossRef]
21. Dalila, N.R.; Arshad, M.K.M.; Gopinath, S.C.B.; Norhaimi, W.M.W.; Fathil, M.F.M. Current and future envision on developing biosensors aided by 2D molybdenum disulfide (MoS_2) productions. *Biosens. Bioelectron.* **2019**, *132*, 248–264. [CrossRef] [PubMed]
22. Novoselov, K.S.; Geim, A.K.; Morozov, S.V.; Jiang, D.; Zhang, Y.; Dubonos, S.V.; Grigorieva, I.V.; Firsov, A.A. Electric field effect in atomically thin carbon films. *Science* **2004**, *306*, 666–669. [CrossRef] [PubMed]
23. Lee, C.; Wei, X.; Kysar, J.W.; Hone, J. Measurement of the elastic properties and intrinsic strength of monolayer graphene. *Science* **2008**, *321*, 385–388. [CrossRef] [PubMed]
24. Stoller, M.D.; Park, S.; Zhu, Y.; An, J.; Ruoff, R.S. Graphene-based ultracapacitors. *Nano Lett.* **2008**, *8*, 3498–3502. [CrossRef] [PubMed]
25. Ghosh, S.; Calizo, I.; Teweldebrhan, D.; Pokatilov, E.P.; Nika, D.L.; Balandin, A.A.; Bao, W.; Miao, F.; Lau, C.N. Extremely high thermal conductivity of graphene: Prospects for thermal management applications in nanoelectronic circuits. *Appl. Phys. Lett.* **2008**, *92*, 151911. [CrossRef]
26. Geim, A.K. Graphene: Status and prospects. *Science* **2009**, *324*, 1530–1534. [CrossRef] [PubMed]
27. Ponomarenko, L.A.; Yang, R.; Mohiuddin, T.M.; Katsnelson, M.I.; Novoselov, K.S.; Morozov, S.V.; Zhukov, A.A.; Schedin, F.; Hill, E.W.; Geim, A.K. Effect of a high-kappa environment on charge carrier mobility in graphene. *Phys. Rev. Lett.* **2009**, *102*, 206603. [CrossRef]
28. Ping, J.; Wu, J.; Wang, Y.; Ying, Y. Simultaneous determination of ascorbic acid, dopamine and uric acid using high-performance screen-printed graphene electrode. *Biosens. Bioelectron.* **2012**, *34*, 70–76. [CrossRef]
29. Campuzano, S.; Pedrero, M.; Nikoleli, G.P.; Pingarron, J.M.; Nikolelis, D.P. Hybrid 2D-nanomaterials-based electrochemical immunosensing strategies for clinical biomarkers determination. *Biosens. Bioelectron.* **2017**, *89*, 269–279. [CrossRef]
30. Stankovich, S.; Dikin, D.A.; Dommett, G.H.B.; Kohlhaas, K.M.; Zimney, E.J.; Stach, E.A.; Piner, R.D.; Nguyen, S.T.; Ruoff, R.S. Graphene-based composite materials. *Nature* **2006**, *442*, 282–286. [CrossRef]
31. Geim, A.K.; Novoselov, K.S. The rise of graphene. *Nat. Mater.* **2007**, *6*, 183–191. [CrossRef]

32. Li, X.; Cai, W.; An, J.; Kim, S.; Nah, J.; Yang, D.; Piner, R.; Velamakanni, A.; Jung, I.; Tutuc, E.; et al. Large-area synthesis of high-quality and uniform graphene films on copper foils. *Science* **2009**, *324*, 1312–1314. [CrossRef] [PubMed]
33. Reina, A.; Jia, X.; Ho, J.; Nezich, D.; Son, H.; Bulovic, V.; Dresselhaus, M.S.; Kong, J. Large area, few-layer graphene films on arbitrary substrates by chemical vapor deposition. *Nano Lett.* **2009**, *9*, 30–35. [CrossRef] [PubMed]
34. Chhowalla, M.; Shin, H.S.; Eda, G.; Li, L.J.; Loh, K.P.; Zhang, H. The chemistry of two-dimensional layered transition metal dichalcogenide nanosheets. *Nat. Chem.* **2013**, *5*, 263–275. [CrossRef] [PubMed]
35. Georgakilas, V.; Otyepka, M.; Bourlinos, A.B.; Chandra, V.; Kim, N.; Kemp, K.C.; Hobza, P.; Zboril, R.; Kim, K.S. Functionalization of graphene: Covalent and non-covalent approaches, derivatives and applications. *Chem. Rev.* **2012**, *112*, 6156–6214. [CrossRef] [PubMed]
36. Kelly, K.F.; Billups, E. Synthesis of soluble graphite and graphene. *Acc. Chem. Res.* **2013**, *46*, 4–13. [CrossRef]
37. Xu, M.; Liang, T.; Shi, M.; Chen, H. Graphene-like two-dimensional materials. *Chem. Rev.* **2013**, *113*, 3766–3798. [CrossRef]
38. Ciesielski, A.; Samori, P. Graphene via sonication assisted liquid-phase exfoliation. *Chem. Soc. Rev.* **2014**, *43*, 381–398. [CrossRef]
39. Randviir, E.P.; Brownson, D.A.C.; Banks, C.E. A decade of graphene research: Production, applications and outlook. *Mater. Today* **2014**, *17*, 426–432. [CrossRef]
40. Duong, D.L.; Yun, S.J.; Lee, Y.H. van der Waals layered materials: Opportunities and challenges. *ACS Nano* **2017**, *11*, 11803–11830. [CrossRef]
41. Manzeli, S.; Ovchinnikov, D.; Pasquier, D.; Yazyev, O.V.; Kis, A. 2D transition metal dichalcogenides. *Nat. Rev. Mater.* **2017**, *2*, 17033. [CrossRef]
42. Yu, X.; Cheng, H.; Zhang, M.; Zhao, Y.; Qu, L.; Shi, G. Graphene-based smart materials. *Nat. Rev. Mater.* **2017**, *2*, 0014. [CrossRef]
43. Li, H.; Li, Y.; Aljarb, A.; Shi, Y.; Li, L.J. Epitaxial growth of two-dimensional layered transition-metal dichalcogenides: Growth mechanism, controllability, and scalability. *Chem. Rev.* **2018**, *118*, 6134–6150. [CrossRef] [PubMed]
44. Ambrosi, A.; Chua, C.K.; Bonanni, A.; Pumera, M. Electrochemistry of graphene and related materials. *Chem. Rev.* **2014**, *114*, 7150–7188. [CrossRef]
45. Wang, L.; Xiong, Q.; Xiao, F.; Duan, H. 2D nanomaterials based electrochemical biosensors for cancer diagnosis. *Biosens. Bioelectron.* **2017**, *89*, 136–151. [CrossRef]
46. Chimene, D.; Alge, D.L.; Gaharwar, A.K. Two-dimensional nanomaterials for biomedical applications: Emerging trends and future prospects. *Adv. Mater.* **2015**, *27*, 7261–7284. [CrossRef]
47. Lawn, J.E.; Cousens, S.; Zupan, J. 4 million neonatal deaths: When? where? why? *Lancet* **2005**, *365*, 891–900. [CrossRef]
48. Fowler, C.C.; Chang, S.J.; Gao, X.; Geiger, T.; Stack, G.; Galan, J.E. Emerging insights into the biology of typhoid toxin. *Curr. Opin. Microbiol.* **2017**, *35*, 70–77. [CrossRef]
49. Blanco, J. Accumulation of dinophysis toxins in bivalve molluscs. *Toxins* **2018**, *10*, 453. [CrossRef]
50. Pitois, F.; Fastner, J.; Pagotto, C.; Dechesne, M. Multi-toxin occurrences in ten french water resource reservoirs. *Toxins* **2018**, *10*, 283. [CrossRef]
51. Janik, E.; Ceremuga, M.; Saluk-Bijak, J.; Bijak, M. Biological toxins as the potential tools for bioterrorism. *Int. J. Mol. Sci.* **2019**, *20*, 1181. [CrossRef]
52. Marchese, S.; Polo, A.; Ariano, A.; Velotto, S.; Costantini, S.; Severino, L. Aflatoxin B1 and M1: Biological properties and their involvement in cancer development. *Toxins* **2018**, *10*, 214. [CrossRef]
53. Partida-Martinez, L.P.; Hertweck, C. Pathogenic fungus harbours endosymbiotic bacteria for toxin production. *Nature* **2005**, *437*, 884–888. [CrossRef]
54. Koropatnick, T.A.; Engle, J.T.; Apicella, M.A.; Stabb, E.V.; Goldman, W.E.; McFall-Ngai, M.J. Microbial factor-mediated development in a host-bacterial mutualism. *Science* **2004**, *306*, 1186–1188. [CrossRef]
55. Richardson, S.D. Water analysis: Emerging contaminants and current issues. *Anal. Chem.* **2007**, *79*, 4295–4323. [CrossRef]
56. Berthiller, F.; Schuhmacher, R.; Adam, G.; Krska, R. Formation, determination and significance of masked and other conjugated mycotoxins. *Anal. Bioanal. Chem.* **2009**, *395*, 1243–1252. [CrossRef]

57. Garcia-Galan, M.J.; Diaz-Cruz, M.S.; Barcelo, D. Combining chemical analysis and ecotoxicity to determine environmental exposure and to assess risk from sulfonamides. *TrAC Trends Anal. Chem.* **2009**, *28*, 804–819. [CrossRef]
58. Man, Y.; Liang, G.; Li, A.; Pan, L. Analytical methods for the determination of alternaria mycotoxins. *Chromatographia* **2017**, *80*, 9–22. [CrossRef]
59. Steil, D.; Pohlentz, G.; Legros, N.; Mormann, M.; Mellmann, A.; Karch, H.; Muething, J. Combining mass spectrometry, surface acoustic wave interaction analysis, and cell viability assays for characterization of shiga toxin subtypes of pathogenic *Escherichia coil* bacteria. *Anal. Chem.* **2018**, *90*, 8989–8997. [CrossRef]
60. Patel, P.D. (Bio)sensors for measurement of analytes implicated in food safety: A review. *TrAC Trends Anal. Chem.* **2002**, *21*, 96–115. [CrossRef]
61. Van Dorst, B.; Mehta, J.; Bekaert, K.; Rouah-Martin, E.; De Coen, W.; Dubruel, P.; Blust, R.; Robbens, J. Recent advances in recognition elements of food and environmental biosensors: A review. *Biosens. Bioelectron.* **2010**, *26*, 1178–1194. [CrossRef] [PubMed]
62. McGrath, T.F.; Elliott, C.T.; Fodey, T.L. Biosensors for the analysis of microbiological and chemical contaminants in food. *Anal. Bioanal. Chem.* **2012**, *403*, 75–92. [CrossRef] [PubMed]
63. Zhang, Z.; Yu, L.; Xu, L.; Hu, X.; Li, P.; Zhang, Q.; Ding, X.; Feng, X. Biotoxin sensing in food and environment via microchip. *Electrophoresis* **2014**, *35*, 1547–1559. [CrossRef] [PubMed]
64. Duracova, M.; Klimentova, J.; Fucikova, A.; Dresler, J. Proteomic methods of detection and quantification of protein toxins. *Toxins* **2018**, *10*, 99. [CrossRef]
65. Peltomaa, R.; Benito-Pena, E.; Moreno-Bondi, M.C. Bioinspired recognition elements for mycotoxin sensors. *Anal. Bioanal. Chem.* **2018**, *410*, 747–771. [CrossRef]
66. Mustafa, F.; Andreescu, S. Chemical and biological sensors for food-quality monitoring and smart packaging. *Foods* **2018**, *7*, 168. [CrossRef]
67. Farabullini, F.; Lucarelli, F.; Palchetti, I.; Marrazza, G.; Mascini, M. Disposable electrochemical genosensor for the simultaneous analysis of different bacterial food contaminants. *Biosens. Bioelectron.* **2007**, *22*, 1544–1549. [CrossRef]
68. Bonel, L.; Vidal, J.C.; Duato, P.; Castillo, J.R. An electrochemical competitive biosensor for ochratoxin A based on a DNA biotinylated aptamer. *Biosens. Bioelectron.* **2011**, *26*, 3254–3259. [CrossRef]
69. Ye, Y.; Guo, H.; Sun, X. Recent progress on cell-based biosensors for analysis of food safety and quality control. *Biosens. Bioelectron.* **2019**, *126*, 389–404. [CrossRef]
70. Collins, M.D.; East, A.K. Phylogeny and taxonomy of the food-borne pathogen Clostridium botulinum and its neurotoxins. *J. Appl. Microbiol.* **1998**, *84*, 5–17. [CrossRef]
71. Bigalke, H.; Rummel, A. Medical aspects of toxin weapons. *Toxicology* **2005**, *214*, 210–220. [CrossRef] [PubMed]
72. Rummel, A.; Haefner, K.; Mahrhold, S.; Darashchonak, N.; Holt, M.; Jahn, R.; Beermann, S.; Karnath, T.; Bigalke, H.; Binz, T. Botulinum neurotoxins C, E and F bind gangliosides via a conserved binding site prior to stimulation-dependent uptake with botulinum neurotoxin F utilising the three isoforms of SV2 as second receptor. *J. Neurochem.* **2009**, *110*, 1942–1954. [CrossRef] [PubMed]
73. Eivazzadeh-Keihan, R.; Pashazadeh-Panahi, P.; Baradaran, B.; de la Guardia, M.; Hejazi, M.; Sohrabi, H.; Mokhtarzadeh, A.; Maleki, A. Recent progress in optical and electrochemical biosensors for sensing of Clostridium botulinum neurotoxin. *TrAC Trends Anal. Chem.* **2018**, *103*, 184–197. [CrossRef]
74. Narayanan, J.; Sharma, M.K.; Ponmariappan, S.; Sarita; Shaik, M.; Upadhyay, S. Electrochemical immunosensor for botulinum neurotoxin type-E using covalently ordered graphene nanosheets modified electrodes and gold nanoparticles-enzyme conjugate. *Biosens. Bioelectron.* **2015**, *69*, 249–256. [CrossRef] [PubMed]
75. Chan, C.Y.; Guo, J.; Sun, C.; Tsang, M.K.; Tian, F.; Hao, J.; Chen, S.; Yang, M. A reduced graphene oxide-Au based electrochemical biosensor for ultrasensitive detection of enzymatic activity of botulinum neurotoxin A. *Sens. Actuators B* **2015**, *220*, 131–137. [CrossRef]
76. Afkhami, A.; Hashemi, P.; Bagheri, H.; Salimian, J.; Ahmadi, A.; Madrakian, T. Impedimetic immunosensor for the label-free and direct detection of botulinum neurotoxin serotype A using Au nanoparticles/graphene-chitosan composite. *Biosens. Bioelectron.* **2017**, *93*, 124–131. [CrossRef]
77. Banno, Y.; Kobayashi, T.; Watanabe, K.; Ueno, K.; Nozawa, Y. 2 Toxins (D-1 and D-2) of Clostridium-Difficile causing antibiotic-associated colitis-purification and some characterization. *Biochem. Int.* **1981**, *2*, 629–635.

78. Depitre, C.; Delmee, M.; Avesani, V.; L'Haridon, R.; Roels, A.; Popoff, M.; Corthier, G. Serogroup F strains of Clostridium difficile produce toxin B but not toxin A. *J. Med. Microbiol.* **1993**, *38*, 434–441. [CrossRef]
79. Sambol, S.P.; Merrigan, M.M.; Lyerly, D.; Gerding, D.N.; Johnson, S. Toxin gene analysis of a variant strain of Clostridium difficile that causes human clinical disease. *Infect. Immun.* **2000**, *68*, 5480–5487. [CrossRef]
80. Poutanen, S.M.; Simor, A.E. Clostridium difficile-associated diarrhea in adults. *Can. Med. Assoc. J.* **2004**, *171*, 51–58. [CrossRef]
81. Kuehne, S.A.; Cartman, S.T.; Heap, J.T.; Kelly, M.L.; Cockayne, A.; Minton, N.P. The role of toxin A and toxin B in Clostridium difficile infection. *Nature* **2010**, *467*, 711–713. [CrossRef] [PubMed]
82. Tenover, F.C.; Novak, W.S.; Woods, C.W.; Peterson, L.R.; Davis, T.; Schreckenberger, P.; Fang, F.C.; Dascal, A.; Gerding, D.N.; Nomura, J.H.; et al. Impact of strain type on detection of toxigenic clostridium difficile: Comparison of molecular diagnostic and enzyme immunoassay approaches. *J. Clin. Microbiol.* **2010**, *48*, 3719–3724. [CrossRef] [PubMed]
83. Joshi, L.T.; Mali, B.L.; Geddes, C.D.; Baillie, L. Extraction and sensitive detection of toxins A and B from the human pathogen Clostridium difficile in 40 s using microwave-accelerated metal-enhanced fluorescence. *PLoS ONE* **2014**, *9*, e104334. [CrossRef] [PubMed]
84. Fang, Y.S.; Chen, S.Y.; Huang, X.J.; Wang, L.S.; Wang, H.Y.; Wang, J.F. Simple approach for ultrasensitive electrochemical immunoassay of Clostridium difficile toxin B detection. *Biosens. Bioelectron.* **2014**, *53*, 238–244. [CrossRef] [PubMed]
85. Liu, M.; Chang, D.; Li, Y. Discovery and biosensing applications of diverse RNA-cleaving DNAzymes. *Acc. Chem. Res.* **2017**, *50*, 2273–2283. [CrossRef] [PubMed]
86. Marrack, P.; Kappler, J. The staphylococcal enterotoxins and their relatives. *Science* **1990**, *248*, 705–711. [CrossRef]
87. Carfora, V.; Caprioli, A.; Marri, N.; Sagrafoli, D.; Boselli, C.; Giacinti, G.; Giangolini, G.; Sorbara, L.; Dottarelli, S.; Battisti, A.; et al. Enterotoxin genes, enterotoxin production, and methicillin resistance in Staphylococcus aureus isolated from milk and dairy products in Central Italy. *Int. Dairy J.* **2015**, *42*, 12–15. [CrossRef]
88. Sharma, A.; Rao, V.K.; Kamboj, D.V.; Gaur, R.; Shaik, M.; Shrivastava, A.R. Enzyme free detection of staphylococcal enterotoxin B (SEB) using ferrocene carboxylic acid labeled monoclonal antibodies: An electrochemical approach. *New J. Chem.* **2016**, *40*, 8334–8341. [CrossRef]
89. Rapini, R.; Marrazza, G. Electrochemical aptasensors for contaminants detection in food and environment: Recent advances. *Bioelectrochemistry* **2017**, *118*, 47–61. [CrossRef]
90. Farka, Z.; Jurik, T.; Kovar, D.; Trnkova, L.; Skladal, P. Nanoparticle-based immunochemical biosensors and assays: Recent advances and challenges. *Chem. Rev.* **2017**, *117*, 9973–10042. [CrossRef]
91. Nodoushan, S.M.; Nasirizadeh, N.; Amani, J.; Halabian, R.; Fooladi, A.A.I. An electrochemical aptasensor for staphylococcal enterotoxin B detection based on reduced graphene oxide and gold nano-urchins. *Biosens. Bioelectron.* **2019**, *127*, 221–228. [CrossRef] [PubMed]
92. Gourama, H.; Bullerman, L.B. Aspergillus flavus and Aspergillus parasiticus: Aflatoxigenic fungi of concern in foods and feeds: A review. *J. Food Prot.* **1995**, *58*, 1395–1404. [CrossRef] [PubMed]
93. Bennett, J.W.; Klich, M. Mycotoxins. *Clin. Microbiol. Rev.* **2003**, *16*, 497–516. [CrossRef] [PubMed]
94. Keller, N.P.; Turner, G.; Bennett, J.W. Fungal secondary metabolism-From biochemistry to genomics. *Nat. Rev. Microbiol.* **2005**, *3*, 937–947. [CrossRef]
95. Geiser, D.M.; Pitt, J.I.; Taylor, J.W. Cryptic speciation and recombination in the aflatoxin-producing fungus Aspergillus flavus. *Proc. Natl. Acad. Sci. USA* **1998**, *95*, 388–393. [CrossRef]
96. Srivastava, S.; Kumar, V.; Ali, M.A.; Solanki, P.R.; Srivastava, A.; Sumana, G.; Saxena, P.S.; Joshi, A.G.; Malhotra, B.D. Electrophoretically deposited reduced graphene oxide platform for food toxin detection. *Nanoscale* **2013**, *5*, 3043–3051. [CrossRef]
97. Srivastava, S.; Ali, M.A.; Umrao, S.; Parashar, U.K.; Srivastava, A.; Sumana, G.; Malhotra, B.D.; Pandey, S.S.; Hayase, S. Graphene oxide-based biosensor for food toxin detection. *Appl. Biochem. Biotechnol.* **2014**, *174*, 960–970. [CrossRef]
98. Srivastava, S.; Abraham, S.; Singh, C.; Ali, M.A.; Srivastava, A.; Sumana, G.; Malhotra, B.D. Protein conjugated carboxylated gold@reduced graphene oxide for aflatoxin B-1 detection. *RSC Adv.* **2014**, *5*, 5406–5414. [CrossRef]

99. Srivastava, S.; Kumar, V.; Arora, K.; Singh, C.; Ali, M.A.; Puri, N.K.; Malhotra, B.D. Antibody conjugated metal nanoparticle decorated graphene sheets for a mycotoxin sensor. *RSC Adv.* **2016**, *6*, 56518–56526. [CrossRef]
100. Zhou, L.; Li, R.; Li, Z.; Xia, Q.; Fang, Y.; Liu, J. An immunosensor for ultrasensitive detection of aflatoxin B-1 with an enhanced electrochemical performance based on graphene/conducting polymer/gold nanoparticles/the ionic liquid composite film on modified gold electrode with electrodeposition. *Sens. Actuators B* **2012**, *174*, 359–365.
101. Shu, J.; Qiu, Z.; Wei, Q.; Zhuang, J.; Tang, D. Cobalt-porphyrin-platinum-functionalized reduced graphene oxide hybrid nanostructures: A novel peroxidase mimetic system for improved electrochemical immunoassay. *Sci. Rep.* **2015**, *5*, 15113. [CrossRef] [PubMed]
102. Wang, D.; Hu, W.; Xiong, Y.; Xu, Y.; Li, C.M. Multifunctionalized reduced graphene oxide-doped polypyrrole/pyrrolepropylic acid nanocomposite impedimetric immunosensor to ultra-sensitively detect small molecular aflatoxin B-1. *Biosens. Bioelectron.* **2015**, *63*, 185–189. [CrossRef] [PubMed]
103. Basu, J.; Datta, S.; RoyChaudhuri, C. A graphene field effect capacitive Immunosensor for sub-femtomolar food toxin detection. *Biosens. Bioelectron.* **2015**, *68*, 544–549. [CrossRef] [PubMed]
104. Hao, N.; Zhang, Y.; Zhong, H.; Zhou, Z.; Hua, R.; Qian, J.; Liu, Q.; Li, H.; Wang, K. Design of a dual channel self-reference photoelectrochemical biosensor. *Anal. Chem.* **2017**, *89*, 10133–10136. [CrossRef] [PubMed]
105. Goud, K.Y.; Hayat, A.; Catanante, G.; Satyanarayana, M.; Gobi, K.V.; Marty, J.L. An electrochemical aptasensor based on functionalized graphene oxide assisted electrocatalytic signal amplification of methylene blue for aflatoxin B1 detection. *Electrochim. Acta* **2017**, *244*, 96–103. [CrossRef]
106. Mo, R.; He, L.; Yan, X.; Su, T.; Zhou, C.; Wang, Z.; Hong, P.; Sun, S.; Li, C. A novel aflatoxin B-1 biosensor based on a porous anodized alumina membrane modified with graphene oxide and an aflatoxin B-1 aptamer. *Electrochem. Commun.* **2018**, *95*, 9–13. [CrossRef]
107. Peng, G.; Li, X.; Cui, F.; Qiu, Q.; Chen, X.; Huang, H. Aflatoxin B1 electrochemical aptasensor based on tetrahedral DNA nanostructures functionalized three dimensionally ordered macroporous MoS_2-AuNPs film. *ACS Appl. Mater. Interfaces* **2018**, *10*, 17551–17559. [CrossRef]
108. Geleta, G.S.; Zhao, Z.; Wang, Z. A novel reduced graphene oxide/molybdenum disulfide/polyaniline nanocomposite-based electrochemical aptasensor for detection of aflatoxin B-1. *Analyst* **2018**, *143*, 1644–1649. [CrossRef]
109. Bhardwaj, H.; Pandey, M.K.; Rajesh; Sumana, G. Electrochemical Aflatoxin B1 immunosensor based on the use of graphene quantum dots and gold nanoparticles. *Microchim. Acta* **2019**, *186*, 592. [CrossRef]
110. Beheshti-Marnani, A.; Hatefi-Mehrjardi, A.; Es'haghi, Z. A sensitive biosensing method for detecting of ultra-trace amounts of AFB1 based on "Aptamer/reduced graphene oxide" nano-bio interaction. *Colloids Surf. B* **2019**, *175*, 98–105. [CrossRef]
111. Tang, Y.; Liu, X.; Zheng, H.; Yang, L.; Li, L.; Zhang, S.; Zhou, Y.; Alwarappan, S. A photoelectrochemical aptasensor for aflatoxin B1 detection based on an energy transfer strategy between Ce-TiO_2@$MoSe_2$ and Au nanoparticles. *Nanoscale* **2019**, *11*, 9115–9124. [CrossRef] [PubMed]
112. Althagafi, I.I.; Ahmed, S.A.; El-Saidid, W.A. Fabrication of gold/graphene nanostructures modified ITO electrode as highly sensitive electrochemical detection of Aflatoxin B1. *PLoS ONE* **2019**, *14*, e0210652. [CrossRef] [PubMed]
113. Zhang, B.; Lu, Y.; Yang, C.; Guo, Q.; Nie, G. Simple "signal-on" photoelectrochemical aptasensor for ultrasensitive detecting AFB1 based on electrochemically reduced graphene oxide/poly(5-formylindole)/Au nanocomposites. *Biosens. Bioelectron.* **2019**, *134*, 42–48. [CrossRef] [PubMed]
114. O'Brien, E.; Dietrich, D.R. Ochratoxin A: The continuing enigma. *Crit. Rev. Toxicol.* **2005**, *35*, 33–60. [CrossRef] [PubMed]
115. Richard, J.L. Some major mycotoxins and their mycotoxicoses-An overview. *Int. J. Food Microbiol.* **2007**, *119*, 3–10. [CrossRef]
116. Pfohl-Leszkowicz, A.; Manderville, R.A. Ochratoxin A: An overview on toxicity and carcinogenicity in animals and humans. *Mol. Nutr. Food Res.* **2007**, *51*, 61–99. [CrossRef]
117. Pfohl-Leszkowicz, A.; Manderville, R.A. An update on direct genotoxicity as a molecular mechanism of ochratoxin A carcinogenicity. *Chem. Res. Toxicol.* **2012**, *25*, 252–262. [CrossRef]

118. Jiang, L.; Qian, J.; Yang, X.; Yan, Y.; Liu, Q.; Wang, K.; Wang, K. Amplified impedimetric aptasensor based on gold nanoparticles covalently bound graphene sheet for the picomolar detection of ochratoxin A. *Anal. Chim. Acta* **2014**, *806*, 128–135. [CrossRef]
119. Qian, J.; Jiang, L.; Yang, X.; Yan, Y.; Mao, H.; Wang, K. Highly sensitive impedimetric aptasensor based on covalent binding of gold nanoparticles on reduced graphene oxide with good dispersity and high density. *Analyst* **2014**, *139*, 5587–5593. [CrossRef]
120. Hao, N.; Jiang, L.; Qian, J.; Wang, K. Ultrasensitive electrochemical Ochratoxin A aptasensor based on CdTe quantum dots functionalized graphene/Au nanocomposites and magnetic separation. *J. Electroanal. Chem.* **2016**, *781*, 332–338. [CrossRef]
121. Bulbul, G.; Hayat, A.; Andreescu, S. A generic amplification strategy for electrochemical aptasensors using a non-enzymatic nanoceria tag. *Nanoscale* **2015**, *7*, 13230–13238. [CrossRef] [PubMed]
122. Loo, A.H.; Bonanni, A.; Pumera, M. Mycotoxin aptasensing amplification by using inherently electroactive graphene-oxide nanoplatelet labels. *Chemelectrochem* **2015**, *2*, 743–747. [CrossRef]
123. Tang, J.; Huang, Y.; Zhang, C.; Liu, H.; Tang, D. Amplified impedimetric immunosensor based on instant catalyst for sensitive determination of ochratoxin A. *Biosens. Bioelectron.* **2016**, *86*, 386–392. [CrossRef] [PubMed]
124. Huang, K.J.; Shuai, H.L.; Chen, Y.X. Layered molybdenum selenide stacking flower-like nanostructure coupled with guanine-rich DNA sequence for ultrasensitive ochratoxin A aptasensor application. *Sens. Actuators B* **2016**, *225*, 391–397. [CrossRef]
125. Sun, A.L.; Zhang, Y.F.; Sun, G.P.; Wang, X.N.; Tang, D. Homogeneous electrochemical detection of ochratoxin A in foodstuff using aptamer-graphene oxide nanosheets and DNase I-based target recycling reaction. *Biosens. Bioelectron.* **2017**, *89*, 659–665. [CrossRef]
126. Koh, V.; Ang, W.L.; Bonanni, A. The Role of Surface Chemistry in Impedimetric Aptasensing. *Chemelectrochem* **2018**, *5*, 3654–3659. [CrossRef]
127. Zhu, X.; Kou, F.; Xu, H.; Han, Y.; Yang, G.; Huang, X.; Chen, W.; Chi, Y.; Lin, Z. Label-free ochratoxin A electrochemical aptasensor based on target-induced noncovalent assembly of peroxidase-like graphitic carbon nitride nanosheet. *Sens. Actuators B* **2018**, *270*, 263–269. [CrossRef]
128. Wang, Y.; Ning, G.; Bi, H.; Wu, Y.; Liu, G.; Zhao, Y. A novel ratiometric electrochemical assay for ochratoxin A coupling Au nanoparticles decorated MoS_2 nanosheets with aptamer. *Electrochim. Acta* **2018**, *285*, 120–127. [CrossRef]
129. Tang, J.; Xiong, P.; Cheng, Y.; Chen, Y.; Peng, S.; Zhu, Z.Q. Enzymatic oxydate-triggered AgNPs etching: A novel signal-on photoelectrochemical immunosensing platform based on Ag@AgCl nanocubes loaded RGO plasmonic heterostructure. *Biosens. Bioelectron.* **2019**, *130*, 125–131. [CrossRef]
130. Kaur, N.; Bharti, A.; Batra, S.; Rana, S.; Rana, S.; Bhalla, A.; Prabhakar, N. An electrochemical aptasensor based on graphene doped chitosan nanocomposites for determination of Ochratoxin A. *Microchem. J.* **2019**, *144*, 102–109. [CrossRef]
131. Sharifuzzaman, M.; Barman, S.C.; Zahed, M.A.; San, N.J.; Park, J.Y. Green synthesis of reduced graphene oxide decorated with few-layered MoS_2-nanoroses and Au/Pd/Ag trimetallic nanoparticles for ultrasensitive label-free immunosensing platforms. *J. Electrochem. Soc.* **2019**, *166*, B249–B257. [CrossRef]
132. Shi, Z.Y.; Zheng, Y.T.; Zhang, H.B.; He, C.H.; Wu, W.D.; Zhang, H.B. DNA Electrochemical aptasensor for detecting fumonisins B1 based on graphene and thionine nanocomposite. *Electroanalysis* **2015**, *27*, 1097–1103. [CrossRef]
133. Chen, X.; Huang, Y.; Ma, X.; Jia, F.; Guo, X.; Wang, Z. Impedimetric aptamer-based determination of the mold toxin fumonisin B1. *Microchim. Acta* **2015**, *182*, 1709–1714. [CrossRef]
134. Lu, L.; Seenivasan, R.; Wang, Y.C.; Yu, J.H.; Gunasekaran, S. An electrochemical immunosensor for rapid and sensitive detection of mycotoxins fumonisin B1 and deoxynivalenol. *Electrochim. Acta* **2016**, *213*, 89–97. [CrossRef]
135. Feng, R.; Zhang, Y.; Ma, H.; Wu, D.; Fan, H.; Wang, H.; Li, H.; Du, B.; Wei, Q. Ultrasensitive non-enzymatic and non-mediator electrochemical biosensor using nitrogen-doped graphene sheets for signal amplification and nanoporous alloy as carrier. *Electrochim. Acta* **2013**, *97*, 105–111. [CrossRef]
136. Jiang, K.; Nie, D.; Huang, Q.; Fan, K.; Tang, Z.; Wu, Y.; Han, Z. Thin-layer MoS_2 and thionin composite-based electrochemical sensing platform for rapid and sensitive detection of zearalenone in human biofluids. *Biosens. Bioelectron.* **2019**, *130*, 322–329. [CrossRef]

137. He, B.; Yan, X. An amperometric zearalenone aptasensor based on signal amplification by using a composite prepared from porous platinum nanotubes, gold nanoparticles and thionine-labelled graphene oxide. *Microchim. Acta* **2019**, *186*, 383. [CrossRef]
138. Dawson, R.M. The toxicology of microcystins. *Toxicon* **1998**, *36*, 953–962. [CrossRef]
139. Landsberg, J.H. The effects of harmful algal blooms on aquatic organisms. *Rev. Fish. Sci.* **2002**, *10*, 113–390. [CrossRef]
140. Codd, G.A.; Morrison, L.F.; Metcalf, J.S. Cyanobacterial toxins: Risk management for health protection. *Toxicol. Appl. Pharmacol.* **2005**, *203*, 264–272. [CrossRef]
141. Chen, L.; Chen, J.; Zhang, X.; Xie, P. A review of reproductive toxicity of microcystins. *J. Hazard. Mater.* **2016**, *301*, 381–399. [CrossRef] [PubMed]
142. Preece, E.P.; Hardy, F.J.; Moore, B.C.; Bryan, M. A review of microcystin detections in estuarine and marine waters: Environmental implications and human health risk. *Harmful Algae* **2017**, *61*, 31–45. [CrossRef]
143. Zhang, W.; Dixon, M.B.; Saint, C.; Teng, K.S.; Furumai, H. Electrochemical biosensing of Algal toxins in water: The current state-of-the-art. *ACS Sens.* **2018**, *3*, 1233–1245. [CrossRef] [PubMed]
144. Vogiazi, V.; de la Cruz, A.; Mishra, S.; Shanov, V.; Heineman, W.R.; Dionysiou, D.D. A comprehensive review: Development of electrochemical biosensors for detection of cyanotoxins in freshwater. *ACS Sens.* **2019**, *4*, 1151–1173. [CrossRef] [PubMed]
145. Wei, Q.; Zhao, Y.; Du, B.; Wu, D.; Cai, Y.; Mao, K.; Li, H.; Xu, C. Nanoporous PtRu alloy enhanced nonenzymatic immunosensor for ultrasensitive detection of microcystin-LR. *Adv. Funct. Mater.* **2011**, *21*, 4193–4198. [CrossRef]
146. Zhao, H.; Tian, J.; Quan, X. A graphene and multienzyme functionalized carbon nanosphere-based electrochemical immunosensor for microcystin-LR detection. *Colloids Surf. B* **2013**, *103*, 38–44. [CrossRef]
147. Li, R.; Xia, Q.; Li, Z.; Sun, X.; Liu, J. Electrochemical immunosensor for ultrasensitive detection of microcystin-LR based on graphene-gold nanocomposite/functional conducting polymer/gold nanoparticle/ionic liquid composite film with electrodeposition. *Biosens. Bioelectron.* **2013**, *44*, 235–240.
148. Du, H.L.; Fu, X.W.; Wen, Y.P.; Qiu, Z.J.; Xiong, L.M.; Hong, N.Z.; Yang, Y.H. A label-free immunosensor for microcystins-LR based on graphene and gold Nanocage. *Chin. J. Anal. Chem.* **2014**, *42*, 659–665. [CrossRef]
149. Tian, J.; Zhao, H.; Quan, X.; Zhang, Y.; Yu, H.; Chen, S. Fabrication of graphene quantum dots/silicon nanowires nanohybrids for photoelectrochemical detection of microcystin-LR. *Sens. Actuators B* **2014**, *196*, 532–538. [CrossRef]
150. Eissa, S.; Ng, A.; Siaj, M.; Zourob, M. Label-free voltammetric aptasensor for the sensitive detection of microcystin-LR using graphene-modified electrodes. *Anal. Chem.* **2014**, *86*, 7551–7557. [CrossRef]
151. Du, X.; Jiang, D.; Dai, L.; Zhou, L.; Hao, N.; Qian, J.; Qiu, B.; Wang, K. Fabricating photoelectrochemical aptasensor for selectively monitoring microcystin-LR residues in fish based on visible light-responsive BiOBr nanoflakes/N-doped graphene photoelectrode. *Biosens. Bioelectron.* **2016**, *81*, 242–248. [CrossRef] [PubMed]
152. Zhang, W.; Han, C.; Jia, B.; Saint, C.; Nadagouda, M.; Falaras, P.; Sygellou, L.; Vogiazi, V.; Dionysiou, D.D. A 3D graphene-based biosensor as an early microcystin-LR screening tool in sources of drinking water supply. *Electrochim. Acta* **2017**, *236*, 319–327. [CrossRef]
153. He, Z.; Wei, J.; Gan, C.; Liu, W.; Liu, Y. A rolling circle amplification signal-enhanced immunosensor for ultrasensitive microcystin-LR detection based on a magnetic graphene-functionalized electrode. *RSC Adv.* **2017**, *7*, 39906–39913. [CrossRef]
154. Zhang, Y.; Chen, M.; Li, H.; Yan, F.; Pang, P.; Wang, H.; Wu, Z.; Yang, W. A molybdenum disulfide/gold nanorod composite-based electrochemical immunosensor for sensitive and quantitative detection of microcystin-LR in environmental samples. *Sens. Actuators B* **2017**, *244*, 606–615. [CrossRef]
155. Pang, P.; Teng, X.; Chen, M.; Zhang, Y.; Wang, H.; Yang, C.; Yang, W.; Barrow, C.J. Ultrasensitive enzyme-free electrochemical immunosensor for microcystin-LR using molybdenum disulfide/gold nanoclusters nanocomposites as platform and Au@Pt core-shell nanoparticles as signal enhancer. *Sens. Actuators B* **2018**, *266*, 400–407. [CrossRef]
156. Hu, Y.X.; Shang, F.; Liu, Y.; Wang, S.; Hu, Y.F.; Guo, Z.Y. A label-free electrochemical immunosensor based on multi-functionalized graphene oxide for ultrasensitive detection of microcystin-LR. *Chem. Pap.* **2017**, *72*, 71–79. [CrossRef]

157. Du, X.; Jiang, D.; Li, H.; Hao, N.; You, T.; Wang, K. An intriguing signal-off responsive photoelectrochemical aptasensor for ultrasensitive detection of microcystin-LR and its mechanism study. *Sens. Actuators B* **2018**, *259*, 316–324. [CrossRef]
158. Zhang, W.; Jia, B.; Furumai, H. Fabrication of graphene film composite electrochemical biosensor as a pre-screening algal toxin detection tool in the event of water contamination. *Sci. Rep.* **2018**, *8*, 10686. [CrossRef]
159. Liu, X.; Tang, Y.; Liu, P.; Yang, L.; Li, L.; Zhang, Q.; Zhou, Y.; Khan, M.Z.H. A highly sensitive electrochemical aptasensor for detection of microcystin-LR based on a dual signal amplification strategy. *Analyst* **2019**, *144*, 1671–1678. [CrossRef]
160. Carmichael, W.W. Toxins of Cyanobacteria. *Sci. Am.* **1994**, *270*, 78–86. [CrossRef]
161. van Apeldoorn, M.E.; van Egmond, H.P.; Speijers, G.J.A.; Bakker, G.J.I. Toxins of cyanobacteria. *Mol. Nutr. Food Res.* **2007**, *51*, 7–60. [CrossRef] [PubMed]
162. Westrick, J.A.; Szlag, D.C.; Southwell, B.J.; Sinclair, J. A review of cyanobacteria and cyanotoxins removal/inactivation in drinking water treatment. *Anal. Bioanal. Chem.* **2010**, *397*, 1705–1714. [CrossRef] [PubMed]
163. Moreira, C.; Azevedo, J.; Antunes, A.; Vasconcelos, V. Cylindrospermopsin: Occurrence, methods of detection and toxicology. *J. Appl. Microbiol.* **2013**, *114*, 605–620. [CrossRef] [PubMed]
164. Quiblier, C.; Wood, S.; Echenique-Subiabre, I.; Heath, M.; Villeneuve, A.; Humbert, J.F. A review of current knowledge on toxic benthic freshwater cyanobacteria-Ecology, toxin production and risk management. *Water Res.* **2013**, *47*, 5464–5479.
165. Hinojosa, M.G.; Gutierrez-Praena, D.; Prieto, A.I.; Guzman-Guillen, R.; Jos, A.; Camean, A.M. Neurotoxicity induced by microcystins and cylindrospermopsin: A review. *Sci. Total Environ.* **2019**, *668*, 547–565. [CrossRef]
166. Falconer, I.R.; Humpage, A.R. Health risk assessment of cyanobacterial (blue-green algal) toxins in drinking water. *Int. J. Environ. Res. Public Health* **2005**, *2*, 43–50. [CrossRef]
167. Zhao, Z.; Chen, H.; Ma, L.; Liu, D.; Wang, Z. A label-free electrochemical impedance aptasensor for cylindrospermopsin detection based on thionine-graphene nanocomposites. *Analyst* **2015**, *140*, 5570–5577. [CrossRef]
168. Bratakou, S.; Nikoleli, G.P.; Siontorou, C.G.; Nikolelis, D.P.; Karapetis, S.; Tzamtzis, N. Development of an electrochemical biosensor for the rapid detection of saxitoxin based on air stable lipid films with incorporated anti-STX using graphene electrodes. *Electroanalysis* **2017**, *29*, 990–997. [CrossRef]
169. Tang, J.; Hou, L.; Tang, D.; Zhou, J.; Wang, Z.; Li, J.; Chen, G. Magneto-controlled electrochemical immunoassay of brevetoxin B in seafood based on guanine-functionalized graphene nanoribbons. *Biosens. Bioelectron.* **2012**, *38*, 86–93. [CrossRef]
170. Morabito, S.; Silvestro, S.; Faggio, C. How the marine biotoxins affect human health. *Nat. Prod. Res.* **2018**, *32*, 621–631. [CrossRef]
171. Eissa, S.; Zourob, M. A graphene-based electrochemical competitive immunosensor for the sensitive detection of okadaic acid in shellfish. *Nanoscale* **2012**, *4*, 7593–7599. [CrossRef] [PubMed]
172. Ramalingam, S.; Chand, R.; Singh, C.B.; Singh, A. Phosphorene-gold nanocomposite based microfluidic aptasensor for the detection of okadaic acid. *Biosens. Bioelectron.* **2019**, *135*, 14–21. [CrossRef] [PubMed]

MDPI
St. Alban-Anlage 66
4052 Basel
Switzerland
Tel. +41 61 683 77 34
Fax +41 61 302 89 18
www.mdpi.com

Toxins Editorial Office
E-mail: toxins@mdpi.com
www.mdpi.com/journal/toxins

www.ingramcontent.com/pod-product-compliance
Lightning Source LLC
Chambersburg PA
CBHW062040220926
51850CB00010B/2613
9783039369201